ENCOU
WI
NAT
D0297540

LESLIE BROWN

ENCOUNTERS WITH NATURE

With illustrations by Doris Tischler

Oxford New York Toronto Melbourne
OXFORD UNIVERSITY PRESS
1979

Oxford University Press, Walton Street, Oxford OX2 6DP
OXFORD LONDON GLASGOW
NEW YORK TORONTO MELBOURNE WELLINGTON
KUALA LUMPUR SINGAPORE JAKARTA HONG KONG TOKYO
DELHI BOMBAY CALCUTTA MADRAS KARACHI
IBADAN NAIROBI DAR ES SALAAM CAPE TOWN

British Library Cataloguing in Publication Data

Brown, Leslie
Encounters with nature
1. Animals, Legends and stories of
I. Title
591′.092′4 QL 791 78–40588
ISBN 0–19–217673–0

Printed in Great Britain by
Cox & Wyman Ltd, London, Fakenham and Reading

To Bo and Dan,
good companions in misfortune,
and
to Barbara and Charles
who worried about us (or me)

Preface

On the fourth of February 1977 I flew in a light aircraft from Gode, on the lower Uebi Shebeli River in the Ogaden of Ethiopia, aiming for Addis Ababa via Diredawa. Our pilot was a Dane, Dan Eynhus; and in the back seat of the uncomfortable little Saab was Bo Rindegaard, the Swedish head of our consultant mission to prepare a water master plan for that arid part of the world. Bo and I were both looking forward to leaving Addis for home the next day, he to go skiing with his family, I to join my wife on a holiday in South Africa. For various reasons we had taken off late and had to cut short our flight plan *en route* to Diredawa in order to reach Addis that night.

Approaching the ridge of the Chercher Highlands near Jijiga and Harrar I felt we were too far east. The ground below, to an ecologist's eye, looked all wrong. I spoke to Dan about it; in reply he simply pointed to his compass, which showed we were on course to Diredawa. We should have passed close to a big mountain called Gara Muleta, 3,350 metres (11,000 feet) high. I ought to have checked on this more carefully, as I had climbed it in 1966, and there was an isolated mountain just to our left which might have been it, though it did not look like it to me. One does not argue with the pilot in charge of the aircraft unless one is very sure of one's facts, and the compass did say quite clearly that we were on course. As a former pilot myself I had to accept its ruling.

As we flew on over the ridge, it looked more and more unlikely, and we saw neither Harrar, nor the lakes near it, nor our objective. We did see, far on our right, a small town with white buildings, and it had an airstrip. At about 5.30 p.m., running short of fuel, we were over some very inhospitable-looking jagged hills on the edge of a desert. Diredawa should have been in sight, or at least the Djibouti–Addis railway, but there was no sign of them. 'We'll have to go back to that little city and overnight there,' said Dan. Had we only known, we were as near to Diredawa as we were to this town, Borama; but we did not know. The aircraft had developed a 45° compass error and we were hopelessly lost. To land on an airstrip near a town was obviously preferable to landing in a desert doubtless inhabited by bloodthirsty nomads.

We circled the strip to drive off the camels, and landed. At once uniformed men armed with sub-machine-guns and rifles rushed out; and from the insignia on their hats I realized we were in Somalia, not

Ethiopia. I had been only too right about being too far east. Thinking that, in the situation, my passport might be useful I started to return to the plane to get it. However, when a sub-machine gun was pointed at me in a meaningful manner my keenness to obtain this vital document gave way to an inclination to co-operate. I went along, as they say.

We were stripped of all our possessions and questioned for some time and were assured that we would not be ill-treated in Somalia – 'Somalis are not like Ethiopians'. We were given a meal of eggs and bread in a local hotel, much cleaner than its counterpart in Ethiopia would have been, then taken to a rest house where people who spoke good English discussed various matters with us. They said they were teachers, but probably they were the local K.G.B. They were rather mystified when I gave them a lecture on the length of the Somali femur as an adaptation to desert life. Finally, late at night we were taken to Hargeisa, where, after further questioning, we slept in the Police Hospital. As I had anticipated spending the night on the concrete floor of a jail, chained to the wall (as would certainly have happened in Ethiopia), the clean and comfortable bugless bed, warm blankets, a shower that worked, and a bearable latrine seemed relative and unexpected bliss.

For the next two days we were questioned by one lot of police after another. We immediately asked to be allowed to send a telegram or to telephone our relatives to say we were safe and being well treated. They would not allow this, and said they would do it. They never did; in fact they deliberately suppressed the news of our landing in order to embarrass the Ethiopians and force them to mount an air search for us, which they did. Of course, no one found us; and for five days our wives believed us dead. There are people who actually like Somalis, but I am not among their number, nor will be in the foreseeable future. My brother, one-time District Commissioner in the northern parts of Kenya, used to say you could respect them to a degree because they did not have to be taught to lie – they knew how to do it from birth. Too true. They lie for the sake of lying, automatically, instantly, and fluently.

We slept in the hospital, and were taken three times daily to the Hargeisa club for meals by a regular policeman, Captain Mohamed Hassan, a jolly type, though he said he was a communist. Then we fell into the hands of a young man called Ahmed, said to be deputy head of the Security Police, and of his immediate superior. Ahmed was unctuously polite, but neither he nor his superior, a fat man who never looked one in the eye, ever told us the truth at any time. They became our detested gaolers. They said we could write letters (censored, of course), which we did; and they said they would send them to our wives and families. None was ever sent; and our families knew nothing of the situation until they received other letters which we smuggled out.

After a few days they either got tired of transporting us, or else could not spare the transport. We spent the whole day at the club, guarded by a youth known as Mohamed Blue, because of the colour of his shirt. He spoke little English; and chewed *Chat* (*Catha edulis* – a stimulant) all afternoon, belching and spitting at intervals. He became so sozzled with this that if we had been young, strong, and had a ready means of escape to Ethiopia we would have snapped his neck and gone. It would have been child's play, and certainly no loss to humanity. But I am now old and infirm, and we hoped too that by co-operating we would soon be freed. That is what they told us, anyway.

We did the things that prisoners do. Whittled a set of chessmen from bits of wood, and played other games, the favourite being a Swedish one in which one throws flat stones to land as near a wooden peg as possible. Bo was too good at chess for me, and too athletic; but I regularly beat Dan at both games, which was boring for him though he played like a good sport all through.

A Swedish diplomat was sent from Addis Ababa to Mogadishu to negotiate our release. The British could do nothing useful. My two obviously Scandinavian companions were dubbed Israeli spies, but no one was in any doubt that I was a hard-boiled ex-British Colonialist, the sort of swine who used to stamp on their faces, and a very dangerous one too. 'Look at all those Ethiopian visas in his passport!' While leaving them in very little doubt as to what I thought of them, I managed to avoid losing my temper and so making things worse for all of us.

After about ten days we were in regular telephone contact with the Swedish diplomat. However, we were not allowed to speak to other Europeans in the club, though a number passed through. We obtained books from an Englishman living there, for which we were very grateful. Then we heard that the diplomat had negotiated our release; and a very nice Somali, a shining exception to the general rule, Ali Sheikh Mohamed, a British-trained ex-director of Public Works and now the Scandinavian consul in Mogadishu, came up to Hargeisa to discuss it. He took out the first letters our wives received; and after he had spoken with the authorities our passports, binoculars, and other things that made a difference to life were restored to us. We were moved to a room at the club, where we remained, and restrictions on speaking to other Europeans were lifted. The Swedish diplomat told us on the telephone that he had actually *seen* the order for our release, which had been transmitted to Hargeisa. We have no doubt that it was but that our captors chose to ignore it.

We taxed the Chief of Security Police with this, but he denied it vehemently. Ahmed, too, was obviously furious. 'That will be very impossible!' was all he said when we asked when we should be freed. Then they trumped up new charges against us. The aeroplane in which we had flown was designed for famine relief in Ethiopia, financed by Swedish religious groups. However, its operator, Carl Gustav von Rosen, since killed by Somali guerrillas at Gode, had wanted to give it to the Ethiopians, and had had Ethiopian registration marks painted on it. When told that it must be Swedish-registered, he had not painted out these letters, but merely slapped a plastic sticker over them with Swedish registration letters. All they had to do was to peel off the sticker, and there were the damning Ethiopian registration marks beneath. Of course, they could have found this out in the first twenty-four hours if they had chosen to do so, and perhaps did, but kept it up their sleeves. The hooks on the wings, from which bags of famine relief food were dropped, were now said to be rocket-launchers! No doubt their Soviet advisers told them how absurd they were; but it did not suit them to let us go. They liked having us in their power.

We were now cut off again from contact with the Swedish diplomat and once more not allowed to speak to Europeans in the club. Of course, we sometimes did, when Mohamed Blue was intoxicated. If we all went in different directions he could not watch all three of us, and scarcely tried. For some reason, my binoculars were not taken away again, and I was able to watch birds. In the month we were there I recorded fifty-two species in the gardens of the Hargeisa club, one of them Wahlberg's Eagle, a new record for Somalia.

We were again questioned repeatedly, in my case late one evening after dinner. No doubt they hoped to force some new and damning admission from me, and had learned such methods in Moscow. I simply told them not to be silly, and repeated what I had said before. Then a senior commissioner was sent from Mogadishu to go into our case in detail. It so happened that a British diver, working on a pipeline in Berbera harbour, had passed through the club a day or so before, and we had smuggled letters out with him. He was arrested on the airport with these; and when I was interrogated by the Commissioner, a very reasonable man, I thought, a grinning Ahmed produced my own letters for me to read. He thought he was dealing a trump card no doubt; but his face fell as I read, for the letters contained such phrases as 'We have been told one string of lies after another' and 'These people obviously hate our guts, especially those of the British'. The Commissioner evidently did not like it; it was not the sort of impression he thought would do Somalia any good. We saw no more of Ahmed for some time, and when we saw him he was apparently in bad odour. We hoped he might find himself in the local variation of Siberia.

We settled into a routine. 'You people are too impatient; you could be here for months!' they said. We played many games of chess, and each morning and evening had a stone-throwing session. I also watched birds morning and evening, drafted my technical report, and then, since I had pen and paper and time to kill, began writing these tales about animals and birds. We were all rather miserable and apprehensive, though not ill-treated. So I thought to beguile the time by recalling golden moments in my life, on the principle of the sundial – 'Counte ye onlie the bright hours' – or whatever it says. I found myself getting more and more involved in this as time went on, and tried out some of the stories on my companions. Though neither of them was a naturalist, they were taken by the idea. They liked the stories, which I read aloud as we were resting in the afternoon after lunch.

Each morning I wrote, in a thatched pavilion outside the club, where once British officials sat and enjoyed their gin or beer. I took notice of the birds that passed, and in the evening had a proper bird-watching session. I watched some birds more intensively than ever before and noticed some new things about them. Some I had never seen before, including a Sprosser or Thrush-nightingale, not normally my sort of bird. It was growing warmer, and male Vitelline Masked Weavers were active building new nests. I planned a long-term project, to study them through the breeding season, if things went on the same way. There seemed no reason why they should not. We had three square meals a day, and even a free ration of beer or whisky in the evening. Somalis, Muslims all, came and drank hypocritically in the club each evening. The accumulated pile of beer bottles was periodically cleared away by a contractor.

We regarded Ahmed as the *fons et origo* of our protracted detention. We found a large oval stone, about the size of a human head, with depressions for eyes and mouth. We placed it outside the room in which we were living and regularly and ceremonially urinated on it each night until all the surrounding vegetation had died. It was my idea, needless to say, and I intend that Ahmed shall some day learn that his symbolic face was solemnly and contemptuously pissed upon each night by those he held in his power. I was a little surprised by the keenness in this respect shown by my Scandinavian companions. In our loathing of Ahmed we were at one, as in many other ways. Captivity brings out the best, most unselfish instincts in many men.

They were building a night club, destined no doubt to be a superior sort of brothel, next door to our room. And making a real old botch of it too, with an enormous unnecessary foundation trench, and hordes of screaming labourers pouring badly mixed concrete daily. Mercifully they went away at night; but the availability of cement gave us an idea.

After about three weeks at the club, we noticed that the Chief of Security Police was sidling about, trying to speak to us. We pointedly ignored him. Eventually he succeeded in choking back his xenophobia to tell us we should soon be freed. It evidently gave him no pleasure to have to do it. On 9 March Ali Sheikh Mohamed arrived to take us away; and when it was finally certain that we would be freed, there was general jollification all round. The club servants, some of them dating from British times, had always been very cordial to us, and we felt they were relieved rather than otherwise.

Before we left for the airport on the morning of the tenth we took Ahmed's head, the ceremonial oval stone, and cemented it outside the entrance to the club, where any passing dog would be inclined to relieve himself on it. Bo took great trouble over the mixture; and when he was satisfied, wiped his hands and said, 'They won't get that off without a sledgehammer!' So, master Ahmed, if you wonder what that stone was put there for, that is it! There *were* dogs; and we hope that many have saluted the stone by now.

When we eventually left I had written two-thirds of this book, and thought about the rest. It was written to kill time, and to recall pleasant things and places while enduring, if not real hardship, worry and irritation, with the feeling that things might get worse at any moment, especially for me, being British, and not even thought to be only an Israeli spy. The stories helped to pass the time in the afternoons for my supposed Jewish companions in crime. I am proud of the fact that in our time of trial we all behaved like gentlemen and that, despite living in close proximity under somewhat difficult conditions, there was never a cross word between us.

Contents

	Threads of experience	1
1	The Antbear	11
2	Badgers	19
3	Beavers	27
4	Chimpanzees	37
5	Eagle Hill	45
6	Flamingos	67
7	The Honey Badger	91
8	The Mountain Nyala	101
9	Nightjars at Night	127
10	Otters	139
11	Pelican Island	149
12	Tigers	171
13	The Whale Shark	183
	Further Reading	189
	Index	192

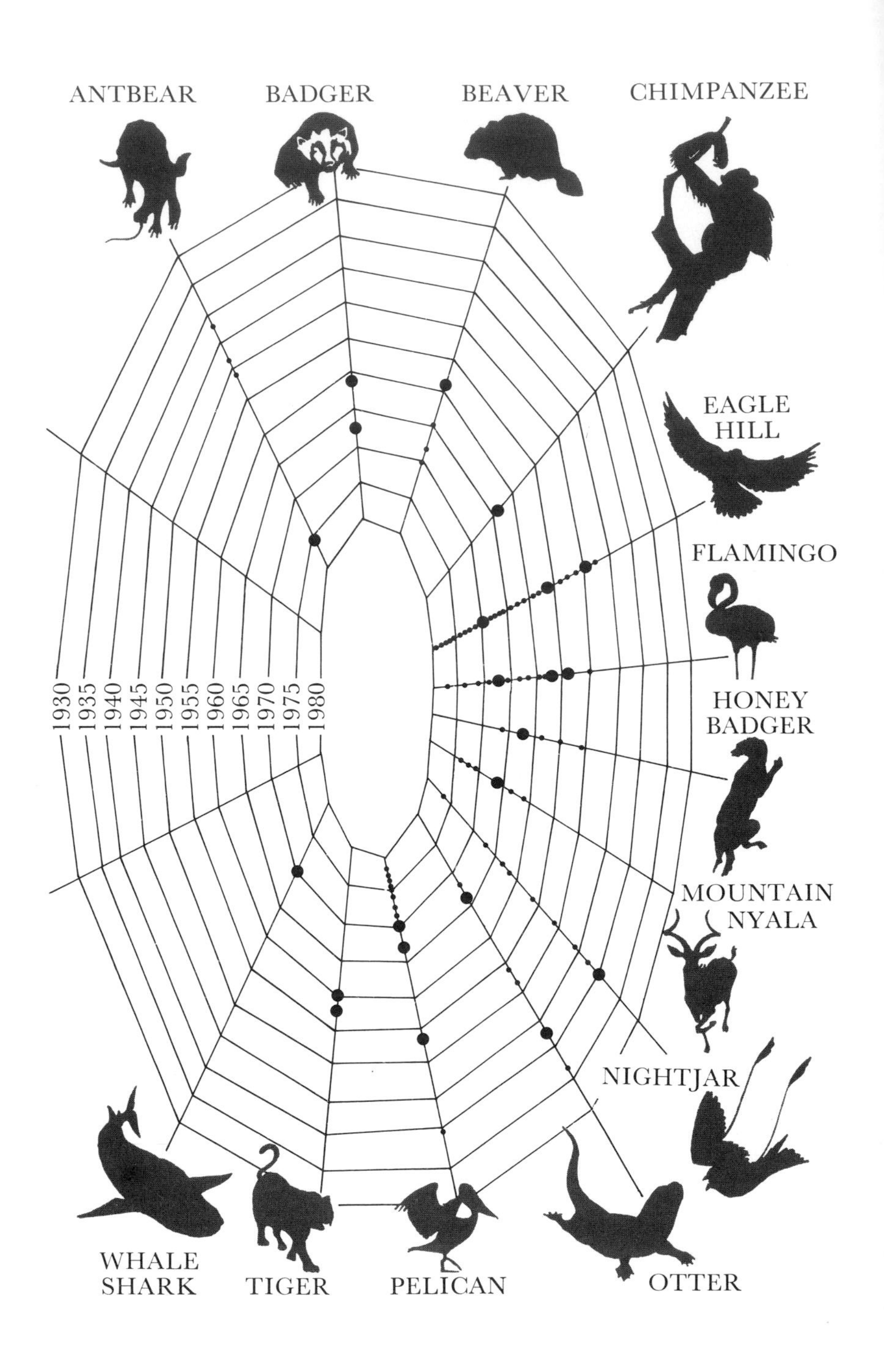
ANTBEAR
BADGER
BEAVER
CHIMPANZEE
EAGLE HILL
FLAMINGO
HONEY BADGER
MOUNTAIN NYALA
NIGHTJAR
OTTER
PELICAN
TIGER
WHALE SHARK
1930
1935
1940
1945
1950
1955
1960
1965
1970
1975
1980

Threads of experience

'The web of experience is largely of your own weaving,' writes Fraser Darling, concerning life on a little island in *A Naturalist on Rona*, a book written, partly at least, to have that 'something else to do' that set me writing this one. Though he was referring to the close circle of life on a small isolated island, it is just as true in the wider world. Experience is largely brought about by yourself; if you never place yourself in a situation where something unusual could happen, the odds are that it will not happen. It may even come as a very nasty shock then, if it does. However, you may spin part of the web; but fate, or nature, or what you will, must also play its part. You can say to yourself, 'If I had not gone to so-and-so I never would have seen it', and so claim some of the credit. But the other thread has to cross your own to make the web. You were there; but you could not compel an Antbear, or a Honey Badger, or a Tiger, or a Whale Shark to be there too. You just gave them the chance to cross your path.

Any observant naturalist, especially if he lives in little-known parts of the world, has many opportunities, even today, to make discoveries great and small. For that matter, discovery of one sort and another is still possible in civilized Europe, North America, or elsewhere. It is just a bit more difficult, often less spectacular and exciting, perhaps just refinement of a great bulk of earlier work. In the little-known tropics one can often start on a completely clean page.

I often observe that in Africa discovery lies on the doorstep for anyone who chooses to pick it up. Never was this more strikingly demonstrated than in 1976 when, after two heart attacks, I crept out of hospital, not able to totter more than a few metres, though all my active life I have thought little of walking up to 800 kilometres (500 miles) in a month over rough country. I felt pretty despairing, but enforced inactivity was made bearable by the presence of a breeding group of beautiful Red-naped Widow-birds, *Euplectes ardens*, in the long grass near my house. I had made odd notes on them before, but had never watched them in any detail. Now, ambling slowly about, and managing by degrees to climb a short distance up and down a gentle slope, I found that I could delineate the territories of nine separate males. I also found that each could be known by his own colour pattern, noted their displays, and recorded how many females each attracted and how many young they reared.

Two months and a couple of field notebooks later I found, as half-expected, that little had ever been recorded about this Widow-bird. For that matter, little seemed to be known about most species of widow-birds, although they are among the commonest and most spectacular of the weaver-birds so typical of vast tracts of African savannas. Even the dramatic Sakabula, with its black, sixty-centimetre (two-foot) flaunting tail, visible a kilometre off in display, does not seem to have been studied in depth. I had made a considerable advance in knowledge, literally from my doorstep, and when weak as the proverbial kitten.

I have made many other minor and major discoveries, mostly about birds, but some about mammals as well. The most recent of the minor ones were the finding of the hitherto unknown eggs of the endemic Ethiopian White-billed Starling, *Onychognathus albirostris*, and of the nest of the White-winged Dove, *Streptopelia reichenowi*. Both came about just because I happened to be at Makalle in northern Ethiopia and at Gode on the lower Webi Shebeli, doing consultant work, and looked around me as well. More exciting was the moment when, on top of a remote peak in the Bale Mountains of Ethiopia, a characteristic familiar call, 'Kyaw', told me I had extended the range of the Common Chough by about 800 kilometres to the south-east, before I even saw the bird flying past. It just goes to show that, in a place like Ethiopia, all you need to do is to keep eyes and ears open and something is bound to turn up.

Greater discoveries may be planned for, and worked for, over a number of years, or may suddenly burst upon one. My most exciting major discovery was the hitherto unknown breeding ground of the Lesser Flamingo in the middle of Lake Natron. It burst upon me suddenly, not where I thought it might be, and came near to ending my career altogether later, for I tried to reach it over those awful mud-flats, was stuck in the mud in that dreadful place, and escaped alive only with great difficulty and pain. Perhaps even more satisfying in some ways was the long-deferred and long-awaited discovery of a breeding colony of Great White Pelicans on an island in Lake Shala in 1965. In neither case were the habits of these birds entirely unknown; but the subsequent longer studies over years, with my friends Emil Urban and Alan Root (who enjoys the unique distinction of having been bitten by both a puff-adder and a bull hippo and survived, with his happy-go-lucky spirit undimmed) have advanced our knowledge of pelicans and flamingos by giant strides.

I have also made many new finds about the ways of birds of prey, especially eagles, both in my native Scotland and in Africa. They have always been my main life's work and enthusiasm. Since I saw my first Golden Eagle in Scotland in 1937, and was fired by it into permanent orbit, I have studied many other species. I still hope to be able to complete long-term studies of the African Fish Eagle and raise it unchallenged to the status of the world's best-known eagle. My latest studies have been dawn-to-dark watches of one or two pairs, almost impossible to do with other big eagles. Although I think I know them pretty well by now, I still see something new almost every time I go out. And what better subject to choose, when one can often do it best in comfort, reclining on the cushions of a boat, and with an iced beer within reach!

Thus some discoveries are small, can be quickly recorded, and thereafter left for someone else to follow up, though not forgotten. Others can be followed up over many years, sometimes for unexpectedly long periods, such as my own on Eagle Hill in Embu District of Kenya, now in their thirtieth consecutive year. If only I had known, when I began, I would have planned it all better. But in 1952, when I left Embu, it seemed improbable that I should still be studying those eagles in 1978.

Acquisition of new knowledge may cease because of the end of a political era; or never start because of the lack of observant naturalists. Anywhere inhabited by Latin Americans or Italians is axiomatically certain to be little known, just because they are not interested, though this may be changing slowly. Ornithology and the study of large mammals have made great strides in Africa in the last thirty years, and did not entirely halt with political independence in some countries. Equally, some countries, such as Uganda, formerly well known, are now almost impossible to work in. The same cannot be said about India where, in a population of 600 million, only about ten names regularly appear, writing about birds, in the pages of the *Journal of the Bombay Natural History Society*. Large mammals have fared rather better, but partly because of interested grant-aided foreigners. Even that ornithological giant and man before his time, Dr. Salim Ali, has written rather little in recent times, though he has left a vast monument to his industry in many books about birds. Field ornithology, it seems, largely ceased in India with independence, about the end of the egg-collecting era. If I went back to India I would hardly have to step outside to make minor new finds.

Discovery can start with something perfectly humdrum, and lead on to greater things. In his book *Birds over America* Roger Peterson tells the story of an inquiring student who asked his professor what he should work on. 'Here on the lawn,' replied the professor, 'are a flock of Brewer's Blackbirds; they will do nicely for a start.' I do not know if the student took the advice; or, if he did, if he then found that many other members of the family Icteridae had not been studied in detail either. Just so John Crook, having begun with Baya Weaver-birds in India, went on to the much more varied weaver-birds of Africa, and from there to the formulation of far-reaching thoughts about flocking seed-eaters and solitary insect-eaters and their ways of life.

Common, obvious, easily watched animals or birds are often ignored in favour of something rare, apparently more challenging, but which will never produce those masses of statistical data, nowadays so important. David Lack once told me that birds of prey were ornithologically unrewarding because they were generally scarce and elusive. As far as I know no one has yet studied in detail the ubiquitous and cheeky Indian House Crow, though a cripple in a wheel-chair could make a good job of it in the suburbs of any large Indian town. He might just need an urchin or two to climb the trees, to see what was in the nests.

Then there are those personal revelations into the ways of some bird or beast, which one makes for oneself when one happens to be in the right place at the right time, and the other thread crosses one's own. The animal concerned may be well known to others; but seeing is believing and, when one least expects it, one may suddenly have an almost startling insight into the way of life of an animal one knows very little about, just because one is there and watching. I have had such marvellous moments with badgers, tigers, otters, Whale Sharks, and others. Maybe, as I have sometimes done, one can follow it up – as I did my first Scottish Golden Eagle. At other times one knows instinctively that such a moment may never recur and savours it as a gastronome savours an unexpectedly tender steak and a more than passable bottle of wine consumed by chance in an unlikely-looking bistro. Any such revelation is an extremely exciting moment for the watcher concerned, even if he later finds that others have seen much the same thing. Then the two can compare notes and make four – but they will never be exactly the same.

Thus sometimes one must seize the moment, knowing it may never recur. Cameras and other paraphernalia then just become rather a nuisance to me and detract from the full enjoyment of what I see. At other times one can follow up this first revelation in other places and with other species. One still unforgettable experience with nightjars in Trinidad gave me the key as to how to catch them, examine, identify, and release them unharmed. If I hear a new one now, I must go and catch it in the familiar way, in itself an enjoyable adventure in the African night.

I confess that I flinch from any record that means I must kill the animal or bird concerned, in homage to that celebrated dictum 'What's hit is history, what's missed is mystery'. I am glad that in recent years science has become a little more flexible and admits that after all it is difficult for a competent naturalist to mistake an adult Verreaux's Eagle or a bull Greater Kudu. I positively hated having to kill a beautiful male De Brazza's Monkey in Ethiopia in 1969. At the time we thought that this was the first ever recorded with certainty in Ethiopia and that it would strengthen possible links between the forests of West Africa and those of Illubabor. Had I known that a gentleman called Thomas had collected several near the Omo River in 1904 I would not have pulled the trigger, though the record did mean an extension of the known range. I took comfort from the fact that I killed him clean, for I was then a good shot. The missionaries with whom we were staying half-jokingly suggested that we should eat him; and perhaps we should have, to expiate what later seemed a needless crime.

Scientists who make such discoveries are often tongue- or pen-tied about them, disguising their feelings in a cloud of statistical equations and a modern, almost intolerable gobbledygook. It is almost as if they had taken to heart the doubtless apocryphal story of the young Scottish minister. He was called to care for a parish in a remote part of the West Highlands, and had to preach twice each Sunday in churches at opposite ends of a long loch. In summer he walked along the shore, rain or fine; but there came on a long hard winter, and in the end he had to skate the length of the frozen loch. In spring, to his consternation, he was haled before the Kirk Sessions and required to explain his 'maist extraordinar' conduct'. Stammering out something about how he had done his best, he was faced with the damning question 'Ah, but the point is, did ye *enjoy* it?'

Many a time, when making a minor or a major discovery, and especially with those unexpected searchlight revelations into the lives of animals I have met by chance, by being in the right place at the right time, I have been possessed by a sudden flood of joy and excitement. At such moments, one of the great sorrows in my life has been that I cannot spontaneously burst into melodious song. Bubbling with enthusiasm and well-being, when I was younger I sometimes tried. However, when I opened my mouth all that came forth was what Kai Lung aptly calls 'a voice like a multitude of corncrakes calling at variance'. Nowadays it would be coldly, and I fear correctly, called a 'vocalization'. So I have just had to learn to keep quiet and let the excitement fizz and bubble inside me without even vocalizing at all. The sounds are so uncouth that even I cannot bear them on a wild mountain top, when alone.

I feel I may be expressing the feelings of other naturalists, who do not or cannot express their own joy and satisfaction at similar happenings, perhaps for fear of the scathing comments of their peers. I am sure that most other naturalists have shared such joys, even though they may not have expressed them. Just so must Hans Kruuk have felt when he found that the despised hyena was not just the filthy scavenger of legend, but the most effective and terrible predator of the African night. Just so Jane Goodall when she first saw a Chimpanzee select and trim a twig and use it for probing a hole in a termite mound – so joining the very select band of tool-makers, hitherto including only man. And just so many another who never recorded his feelings or even his discoveries in print, or who died before he could.

Victorian explorers, decried today by some people because of their 'purple passages', often made a better attempt to say what they felt in the language of their time. Above all, I wish I had been there to watch the German botanist, Welwitsch, faced with the archaic and extraordinary primitive gymnosperm *Welwitschia mirabilis* in the awful wastes of the Namib Desert, fall to his knees in astonished and grateful acknowledgement of the privilege vouchsafed him. He would have had a bad time with the Kirk Sessions.

Darwin must have felt the same about the Galápagos Islands and their fauna which inspired or crystallized his at that time revolutionary thinking about animal evolution. Although many scientists disguise or suppress what they feel or felt in a dry-as-dust text or a lengthy Ph.D. thesis crammed with statistics, I find it hard to believe that any true naturalist can see or find anything really, radically new to him without feeling the flood of excitement and pleasure in discovery – excitement that may remain as memorable even if he later finds that what he saw is commonplace to another who has specialized in the creature concerned. These rich moments of excitement are the fine bottles of heady wine and classic steaks that come one's way amid a general diet of hard, working tack and plain, often dirty water.

Discovery can be strangled at birth by unperceptive advisers who ought to know better but do not; or rejected because it does not fit a generally accepted theory – as if any one theory could possibly explain every single situation. When I was in my Honours year at St. Andrews University I was given the task of making an ecological survey of the Eden estuary, a tract of cold, wet sand and black, glutinous mud. My mentors concluded that I was lazy, not for the right reasons (for I was), but because, in addition to a heavy load of sieves, specimen jars, formalin, and apparatus, I also took my shotgun to shoot duck. It was not right – I was *enjoying* it!

I thought to counter this silly prejudice by studying the food of estuarine birds which, as bilaterally symmetrical warm-blooded vertebrates, I wrongly accepted as animals, therefore worthy of scientific study. The first few shots demonstrated that Sheldduck ate a small mollusc, *Hydrobia ulvae*, common on the slick green seaweed *Ulva lactuca*. Also that several species of waders ate a small burrowing crustacean, *Corophium volutator*, which my sieve had shown to be very common in the estuarine mud. Hot-foot, I went eagerly to my mentors and said I thought I was on to something new – which in fact I was. I was coldly told that birds were an amateur's group, like butterflies, and that I should stick to Nemertine worms. I am glad I did not and stuck to birds, now recognized as fit subjects for study even by quite serious zoologists.

Thirty years later, almost to the day, I appeared at Aberdeen on a television programme with George Dunnet, now Professor of Zoology at the University there. I was talking about Golden Eagles, he about recent discoveries at the University's research station on the Ythan estuary. I remember feeling sorry for him as he said he had no time to fish – in the Ythan of all places – for sea trout. Amongst other things, waders had been found to eat *Corophium volutator* and Sheldduck *Hydrobia ulvae*. I felt like crying out 'Surprise, surprise!', but contented myself by observing that I had known that since 1937, but never published it.

Self-opinionated editors of scientific journals can also stifle the publication of scientific facts which, if not apparently important in themselves, may fit nicely into a jigsaw laid out by someone else. I would never accuse the late and much lamented Reg Moreau, doyen of African ornithology, and brilliant editor of *Ibis* from 1946 to 1963 of being self-opinionated, though no one was better entitled to his ornithological *amour propre*. Yet 'for reasons of space' he wanted to cut out of my first *Ibis* paper a brief observation of a curious display by a Lesser Honey-guide, with loudly whirring wings. I fought him successfully, for I felt it was new; and this snippet was later fielded by Herbert Friedmann, in his treatise on honey-guides, to fit with another isolated but similar observation made by someone else years before and thousands of kilometres away. Two and two then made more than four. It must be rare, for in thirty years since I have only seen it once again, and heard one other ornithologist describe it to me.

If a scientist should know anything, it is that he is not omniscient, nor ever likely to be, and is not even always right. So to those who advise the young I would say, 'Don't choke him off. Make him sweat through the literature by all means, but he may be right, it may be new.' Scientific editors could remember that space is not everything and that more can often be saved by paring that awful jargonese and by relegating those virtuoso mathematical equations to an appendix in small type than by omitting a passage of solid fact. One should not be afraid to admit that one does not know – and I do freely admit that I cannot always recognize a bird of prey, though everyone expects me to be able to do so at a glance.

Many make new finds that they never record; some because they think it is too trivial to be of interest, others because they cannot put pen to paper easily, yet others because they are too lazy, or even because they do not realize they saw anything interesting at all. Unfortunately, the accumulation of new knowledge proceeds at such a pace today that in order to know for certain that something one has seen is new one must often toil through a huge mass of scientific literature in several languages. Life is often simply too short to do all this. So, much that is new never gets recorded at all, or is recorded by someone else who has more time years later.

In the sixties I had a letter from a policeman posted at Illeret in north-west Kenya, saying that he had found a huge colony of Carmine Bee-eaters on flat ground and asking whether this was unusual. I assured him that it was not only a new breeding record for Kenya, but a most unusual way of nesting for this species. I begged him to publish a short note; but he never did, and in the press of events after 1963 I lost the letter and forgot about it. Then, in 1976, I met Dr. Ann Spoerry, of the Flying Doctor Service, at a cocktail party. She told me how she had had to land her aircraft on only half the strip at Illeret because the rest was covered with nesting Carmine Bee-eaters! Not only did she publish a short note at my persuasion, but she dug up the fact that von Höhnel, almost ninety years earlier, had found Carmine Bee-eaters breeding in flat ground near Illeret. He had not recognized them, but had adequately described them in his journal. Of course, once it had started, others rushed to say that they had known about it for years! Well – they could have said so before.

At the moment I am encouraging Morag Candy, the wife of a tea estate manager at Kaimosi in west Kenya, to become the world's expert on turacos. Turacos are a group of spectacularly beautiful birds peculiar to Africa, but which, curiously, have been little studied. She has Great Blue Turacos and Ross's Turacos breeding in her garden; and as her children have just gone to boarding-school, she lacks that 'something else to do' that Fraser Darling thought so important on Rona. Morag is now firmly in orbit, though I doubt if she has reached her apogee. She will! She already knows more about nesting turacos than anyone else.

So, much discovery, sometimes minor, never gets written up, or is not followed up, or is suppressed by a sticky editor, or gets lost in the past literature and forgotten. Better, more modern, comprehensive reference works would help to remedy this state of affairs. However, with costs what they are publishers often will not produce such books; and so there is nothing for it but a slog through the literature in half a dozen languages.

Sometimes, then, a thing that seems wholly new is not. In the mid-sixties Jane and Hugo Van Lawick saw Egyptian Vultures breaking Ostrich eggs by hurling stones at them. It was hailed as a great new find because, by so doing, the despised scavenging Pharaoh's Chicken joined the select group of tool-users, including man, the higher apes, the Sea

Otter, and a few other birds. Then an industrious literature ferret, Bob Baxter, dug up a nineteenth-century account of Egyptian Vultures breaking Ostrich eggs with stones – though precisely how was not clear. And in 1977 Jeffery Boswall of the BBC Natural History Unit located another nineteenth-century reference, this time an exact account of what the Van Lawicks had seen and thought to be new. It had been lost in the literature – discarded as a traveller's tale; but those Victorian travellers were not such fools as people often make out. Herodotus, still further back, recorded that the Egyptian Plover – his Trochilus – picked the teeth of Nile Crocodiles. I live in hope that someone may yet see one do so again – for in case you are in any doubt no naturalist has, though the tale appears regularly in the writings of novelists who want to seem knowledgeable about the bush.

The more one learns the more there is to learn, and one bit of new knowledge just leads on to another. A new, up-to-date reference book becomes of limited value within a decade, simply because it pinpoints gaps in knowledge, easily gathered fruits. Moreover, anyone can make discoveries; it needs no Ph.D. or even a degree, just a pair of observant eyes and ears and a questing mind. Some of the very best naturalists of recent times have been, scientifically speaking, rank amateurs. Although much natural history literature is dated almost as soon as it appears, it should still be published, for it is the spring-board from which someone else may dive deeper into a new and unknown sea.

I myself hope to spend some of my remaining years floating idly in calm warm seas and watching the behaviour of coral fish, in a kaleidoscopic and infinitely varied world of reefs and lagoons, where the incoming tide will return me gently to my home beach, if I do not fight it like King Canute. *Anno Domini* may make eagles and mountains a little too difficult – but one just needs to change direction. I know quite well that the first day I spend watching in detail the doings of even quite a common butterfly fish such as *Chaetoton trifasciatus* I shall probably learn more about it than anyone else has ever recorded, in sun-spangled milk-warm waters and an incomparably rich environment. I am going to enjoy it, and go on enjoying it as long as I can, and not be ashamed of enjoyment. Maybe, one day, I shall meet a big whale face to face; and it will grin at me and be curious, as I am told they do. That may not be of much interest to the whale expert who shows me the creature – though perhaps I will be lucky and meet one by myself. To me it will be as great a moment as the otter in Loch na Seilg, the Guddesal Tiger, or the unexpected Honey Badger in the wild wastes of the Kalahari Desert.

D. TISCHLER

1. The Antbear

My son Charles and I had gone down into the Kedong Valley from Karen for an afternoon's outing. Our objective was to check a reported nest of Verreaux's Eagle, and to see the effects of three years of drought which had rendered the floor of the Rift Valley in those parts a desert of bare earth, stones, and useless weeds. We were on our way back at about six o'clock of a dull evening, having achieved our objectives. We stopped to look at and count a small herd of impala on the right of the road; and then, there it was!

An Antbear or Aardvark was moving slowly towards us through light bush, fifteen metres away, on the left-hand side of the road. We sat still, thinking it would see us, panic, and immediately bolt. It saw us quite soon; but to our astonishment paid no attention whatever to the car or to us sitting in it. It came on through a gap between the bushes right into the open and then shuffled out into the ditch beside the road, within ten metres of the car. Charles bemoaned the fact that the light was too poor for colour photography; but I was quite content just to sit and watch, for I knew at once that this was a rare occasion to be savoured.

The rather ungainly creature moved slowly along the ditch and roadway, stopping every so often to sniff at the ground. It appeared to be eating single termites, perhaps harvester termites, among the dead vegetation there. Its curious, elongated snout with its flattened tip, vaguely resembling that of a long-nosed pig, searched around and apparently located the insects by smell. Then the tip of the tongue came out and licked up whatever was found. Even with 12-power binoculars I could not definitely see what it was eating, but I guessed it must be termites as they are the Antbear's staple food and some termites are found on the surface in the evening eating dead dry grass stems.

There were no large termite heaps in the area, but at one pile of old road spoil which looked vaguely like a termitarium the Antbear stopped for a minute. 'Cratch—cratch—cratch' went its claws; and in no time it was down fifteen centimetres (six inches) into the brick-hard ground. It blew into the hole and its face disappeared into a cloud of dust. Evidently it found nothing worth having, for it turned and slowly moved back into the bush beside the road. The last we saw of it was its whitish, thickened, trailing tail disappearing between bushes.

We had had it in view for perhaps fifteen minutes; and those fifteen minutes revealed more about Antbears to me than all the rest of the thirty-six years I have been in Africa. The creature seemed in rather poor condition, with its hip-bones showing, which perhaps accounted for the fact that it was so early abroad, well before dark, and may also have been due to the drought. I have otherwise only had brief glimpses of one as it scoots across a road in the headlights of my car. Aardvarks live on or visit my property at Karen; and in the days when we grew coffee they were always disturbing the mulch of dry grass to get at the termites that ate it. There is a burrow or two belonging to these strange beasts in the lower part of my plot, near the river, but I have never actually seen one here, even late in the evening, though I know the burrow is occupied from time to time as the cobwebs at its entrance are cleared. Once, on a moonlit night, I heard an animal, perhaps an Antbear, snuffling outside my bedroom window, but when I rose to see what it was there was nothing but short cropped lawn bathed in the light of the moon.

Thus a common, widespread animal can live close to mankind almost unseen. Hugh Lamprey, for years Director of the Serengeti Research Institute, said, when I told him about it, that he had never seen an Antbear in daylight, though he must have had far better chances than I ever had. It was one of those chance moments of revelation into the life of an animal which may never recur in the lifetime of an ordinary naturalist who does not set out specifically to study only that particular beast. It was the latest of such moments I have had; and although imperfect, tantalizingly short, and impossible to record in photographs, it put me in mind of other such moments with other beasts and birds, and made me think first of writing this book about a naturalist's life and the big and small discoveries that came his way.

It was not, of course, my first actual acquaintance with Antbears or Aardvarks (earth-pigs). These names are misleading as they are actually neither bears nor pigs, nor do they eat ants; but I suppose they get the name 'ant-bear' from their formidable, vaguely bear-like, digging claws, and 'earth-pig' from their vaguely porcine physiognomy. In actual fact they are creatures all on their own, the sole living representative of a mammalian order, the Tubulidentata, so called because of their peculiar teeth, which are almost uniform, numerous, and tube-shaped. The animal needs nothing better to chew up the soft-bodied helpless termites on which it feeds. These are often known as white ants, though not closely related to ants in the insect family tree. There is no really descriptive name for this peculiar, exclusively African creature. *Orycteropus afer*, its scientific name, merely means 'the grey animal that burrows with its feet'.

Aardvarks are distributed throughout African savannas wherever there are termites in sufficient quantity. They are big beasts, so that they need lots of termites, and do not occur where these are very scarce. Thus their main haunts are in the extensive savannas, not in really arid country or dense forest. They must also have soft earth in which to dig their burrows; they cannot do so in rocky ground or in heavy clay which becomes waterlogged during the rains. Apart from that, they are everywhere; and archaic-looking as they are, they are highly specialized for their particular niche in life, and very successful.

They have a humped back, a long head with a flattened tip to the snout, very long trumpet-like ears probably useful both for locating termites and for sensing the approach of danger, and a thick, flattened tail, almost triangular in section, covered with whitish hair, which is supposed to act as a guide to a baby Aardvark following its mother through the bush. Their feet are equipped with huge elliptical toe-nails which leave a distinctive track on soft ground and are used – as I saw – for ripping into the brick-hard earth of termite mounds in a few seconds, and also as weapons of defence. When an Aardvark is attacked it turns over in a lightning somersault and can then strike upwards at the belly of a lion or leopard with those formidable claws. Thus an apparently rather helpless sort of animal is able to live in numbers among powerful, predatory beasts that think little of killing full-grown bull buffaloes.

Aardvarks are very useful to many other animals because of the burrows they make to live in by day. When not used by Aardvarks these are inhabited by wart-hogs in particular, but also by big snakes such as pythons, by jackals, hyenas, wild dogs, and even leopards. An Aardvark burrow inhabited by the owner is usually recognizable by the freshly dug earth outside the entrance, for none of these other animals really digs much. Each Aardvark has a home which consists of a number of burrows, probably near the centre of its territory. At night it emerges and goes the rounds of a series of termite heaps. It harvests the termites in rotation, without finally depleting the essential source of nourishment, by digging a hole and then thrusting its long sticky tongue far into the tunnels of the heap. Termites, hurrying to defend their ravaged home, either stick to the tongue or perhaps (as they do to the probing stick tools used by Chimpanzees) seize on to it with their pincer jaws. The Aardvark draws them out and presumably sucks them off its tongue before repeating the process. It is regarded as a wholly beneficial animal because it eats termites, but it is only one of many creatures to utilize this abundant source of protein and fat and makes little real difference to the total numbers of termites existing. Aardvarks will not prevent your house from falling down; rather, like the old-time rabbit trappers, they ensure that there are enough termites for themselves by harvesting them sensibly; and that means plenty of termites.

Aardvarks can dig their burrows with almost incredible speed. One learns that if one tries to dig one out. Once in Nigeria I tried to do this in wet sandy soil with a gang of tough powerful men with shovels. They laughed at me and said it was a waste of time before I began. After three or four hours they had dug a trench about four and a half metres (fifteen feet) long to a depth at which only their heads were visible. There was still no sign of the owner, although I knew it was inside by the fresh tracks in soft earth after rain that I had found at the mouth of the burrow soon after dawn that morning. I went into the trench and, putting my ear to the end, I fancied that I heard the animal moving within, perhaps digging away in front of us. I gave up as it was obviously useless to continue. My gang just said, 'We told you so.'

Indeed, if it were easy for human beings to dig out or trap Aardvarks they would not be common in densely inhabited countries. Tribal taboos may come into it sometimes: many tribes will not eat animals without definite hooves, but in a country such as Nigeria where people think nothing of eating puff-adders and bush rats, there must have been some other reason why the Aardvark was not often caught and eaten. Nigerian hunters also told me that it was an extremely difficult beast to catch in a trap, even set by expert trappers. Apparently the Aardvark either detects the trap (which it possibly could because of an unusually keen sense of smell) or emerges on the next night from a different entrance to an underground warren. According to the experts who have studied Aardvarks for any length of time, they may have a series of burrows leading to a central chamber, where they sleep all day. Burrow entrances may be thirty metres apart, and the whole complex may have twenty to thirty openings spread over some 420 square metres (500 square yards).

Most of the Aardvark burrows one finds are incomplete, and are not used by the Aardvarks themselves. Hyenas and wild dogs may dig them out and enlarge them, making permanent dens in which they rear their own young, but the animal that uses them most often is the Wart-hog. Wart-hogs escaping from a predator rush up the mouth of an Aardvark burrow, whip round, and back rapidly down, so that the would-be killer then has to face very formidable tusks and not a soft and meaty behind. Wart-hogs can do terrible damage with their lower tusks, which are honed sharp by continual rubbing on the much longer and thicker upper tusks, which look more dangerous than they actually are.

When I was Agricultural Officer in Embu District I had three bull-terriers: Meg, a black-and-tan bitch, and after she died two of her sons. They were all big, powerful dogs, weighing around thirty kilogrammes (seventy pounds) and quite unlike any of the small, distorted, roman-nosed bull-terriers one sees in a show ring nowadays. We used to go hunting together, but they were always almost uncontrollable when they saw an animal they could pursue. Most antelopes they could not catch; but they could catch rhinos, Wart-hogs, and baboons, and these they loved to pursue above all others. The rhinos could do them no harm – they were far too agile; but I had many a desperate half-mile run, yelling and calling them back, to protect them from Wart-hogs and baboons. Sometimes the Wart-hogs escaped down an Aardvark burrow before I could get there, and then I was in real trouble because the dogs went down the burrow too, as far as they could, and tackled the pig face to face in the dark and dust.

Once, when I reached the burrow, I found that an enormous boar had not been able to go very far down. The tails of the two sons, Patchy and Tippy, were protruding from the mouth of the burrow. Peering in and listening to the gruntings and pantings within, I guessed that both dogs had got hold of the pig, and that he had not been able to do much damage so far. In fact they had him on either side of his snout above the tusks. He could not drag them farther in because the burrow was shallow and partly finished, while they could not drag him out as he weighed 82 kilogrammes (180 pounds) and was well jammed. In this impasse there seemed only one thing I could do, which was to get the African I had with me to take hold of the inside back leg of each dog and pull, while I waited with the gun to give the *coup de grâce* as soon as the pig's forehead appeared. He did so, and the whole lot came out like a cork out of a bottle, the bull-terriers stretched taut between their locked jaws and their back legs, the pig unavailingly trying to back down. I shot him as soon as I could, thrusting the barrel of the gun down the hole. The dogs never let go and we dragged out the boar by their hold on his nose.

On another occasion we had run far after the dogs and had finally had to track them to the mouth of an obviously deep burrow. There were snufflings and pantings far within. I had with us a favourite bush companion, Nyagga Mukinyu (which I am told means 'Ostrich, the son of Completed Thing' – though anything human much less complete can hardly be imagined). I suggested kindly that he should go down the burrow and see what he could do, to which he replied 'They're your dogs, you go!'

There are times when it will not do to show how inferior and cowardly one feels, and this was one of them. So I laid down my gun and crawled in. As a result I am able to assert that an Aardvark has a girth closely similar to my own when I was in my prime, weighing about ninety kilogrammes (two hundred pounds), the same as a big boar Wart-hog.

My body very soon shut out all light and most air, and as I moved forward, reaching in front with my hands, I expected to be cut to pieces by a furious pig at any minute. Then my questing hands came in contact with a body, wet with blood, but mercifully covered with long stiff hair – the Wart-hog. In the struggle the dogs had killed it, but they had got jammed beyond it in the burrow so that they would have had to eat their way out; I could hear them panting beyond the body. Reassured in part, I took hold of one of the pig's hind legs and instructed Nyagga to seize mine and pull. He did; and once more the whole caboodle came out like a cork from a bottle. The two naughty dogs who had got me in this literal fix (for I doubt if I could have backed out myself, uphill) emerged grinning, well pleased with themselves, and quite unharmed except for a few small cuts.

Africans eject Wart-hogs from Aardvark burrows by pushing long leafy branches down into their faces till they cannot stand it and bolt. I suppose lions do it by reaching in and hooking their claws in the beast's face and then pulling him out. The story I like best is that one can also eject them by making a crackling noise like a bush fire by crumpling toilet-paper at the entrance. The soft stuff in use nowadays would not do at all; but the stiff Bronco of my youth would do very well. To me this story, perhaps apocryphal, conjures up the picture of a shy hunter in an open grassy plain, seeking a little privacy for natural reasons; and when he had done, being suddenly, unexpectedly, and without doubt unforgettably hoist with his own petard.

Thus the Aardvark has many and varied side-effects on other animals; and it is not because it is uneatable that it is not often eaten by man. A District Officer named Spottiswoode (always known as 'Spode') at Auchi in North Benin once took away from African hunters the Aardvark they had just killed, admonishing them that it was a harmless and beneficial animal that they ought not to persecute. Having done so, and being tired of tough goat, he put it in his refrigerator and ate it by degrees. We had slices with scrambled eggs in the mornings when I was staying with him. It was good, white and close-grained like pork, but stronger-tasting, much better than goat or even chicken. Some otherwise defenceless animals are unpalatable or actually poisonous; not so the Aardvark.

In fact the Aardvark is far from defenceless, though it may look it. Its ears must be exceedingly acute and can probably detect a possible predator from afar. If attacked it cannot run fast, but if cornered it turns a somersault and brings the claws on both fore and hind feet to bear. These claws could evidently disembowel a soft-skinned cat quite easily. Alan Root has a tame Aardvark, which becomes active in the evening, when he plays with it; it rolls on its back but does him no harm with its claws. It is fascinating to feel this nice animal snuffling at one's hand, the tip of its sensitive velvety snout sucking like a little vacuum-cleaner.

Lions probably do sometimes eat Aardvarks, but they keep clear of the claws. I recently saw a dreadful French film about wildlife called *Tooth and Claw* which sickened and bored me with its over-long endless repetition of bloody scenes of lions killing and eating their prey. However, it had some brilliant footage, including shots of a lioness holding an Aardvark by its thickened tail while the poor animal strove to escape by digging its claws into the ground. In the film it was allowed in the end to get away; but one supposes that more than one lioness acting together, as might normally happen, would soon have fixed it. Perhaps lions, unlike Spode, do not like the taste of Aardvark.

Most people never see an Aardvark for more than a few seconds in the light of a car, but their traces are everywhere in the African savannas, even in densely populated cultivated country. Those people who are lucky, perhaps in the late evening, may have such a moment of revelation into its way of life, just because they are there, unexpectedly, at the right time.

D. TISCHLER

2. Badgers

My encounters with British Badgers were a good example of how circumstances may lead to new revelations about the ways of birds and beasts. Most of them took place in Wytham Wood, near Oxford, in 1961. I was at the time a very hard-worked senior officer in the Colonial Service, knowing that the British intended to throw away their Empire. I was poor, because I had sunk all my savings in a large house in Kenya which by 1961 had decreased in value to a fifth of what I had paid for it three years earlier, and my son was a small and expensive baby. I was trying, during four months' leave from my onerous duties, to complete the reading for a standard work on the world's birds of prey, written in collaboration with Dean Amadon of the American Museum, New York. My wife was not with me as she did not want to leave home at that time. I had been to the United States to work with Dean for six weeks, and in the six weeks that remained I was desperately trying to read as much as I could of the available literature on birds of prey. Had I known then that the book would only be published seven years later after many uncertainties and delays I would not have pushed myself so hard; but I did not.

Since I was hard up I was living in a converted ambulance, fitted up as a motor caravan. A ponderous machine, capable at best of 80 kilometres (50 miles) an hour, it was nevertheless commodious and comfortable, with two long bunks, a stove, a sink, and a sort of table bolted to the door on which I could type. Carrying my house with me I called it 'The Snail'; I later passed it on to my brother, who used it for five months of his leave. Since then we have both adopted this way of life whenever we are in Europe alone – as he usually is. We call it 'snailing'; and it is a good way of life, if you are prepared to reduce your wants to simplicities and do not want to bath too often. Dinner in a hotel attends to that need anyway.

In Scotland in those days it was not too hard to find a really agreeable lay-by in which to park for any length of time. Since then a dog-in-the-manger attitude has led to the wiring-off or ditching of many nice spots, though the *cognoscenti* of this mode of travel can still locate disused quarries and the like. In England it is very different. Every wood has notices warning that 'Trespassers will be prosecuted'; gates are locked; and every nice place is fenced or walled in. One grows choosy about such things, and a lay-by beside a main road, plastered with rubbish, and deafening with the din of passing traffic never was for me. Nor are there, in Britain, the really pleasant organized caravan sites one can find on the

Continent. Most sites are in rather dubious fields not useful for anything else; and they also attract persons with gregarious tastes, to whom the blare of the devilish transistor set is actually an attraction, and who do not want to watch squirrels eating mushrooms at dawn – as they do.

Thus, while I was working at the Edward Grey Institute in Oxford, I sought a pleasant spot, not too far out, in which to park regularly. I tried several places without much luck and spent the August Bank Holiday camped in a wood, which shall be nameless, but which I afterwards learned was the sort of place difficult to enter even with a permit. It was securely fenced and most gates were locked. However, on one new gate the hasp was only painted over. It was quite tightly stuck with paint, and might have deterred a law-abiding Englishman, but to a Scot from Kenya, equipped with a tyre lever, it presented no problem. I was inside in five minutes, with the gate closed behind me, and I spent the four days of the holiday typing away in a hidden depression, within a short distance of a keeper's house. Each night I went for a walk when I felt sure that, after feeding his dogs, he would be watching the television. He never knew I was there.

I had lovely walks on the summer nights down the central ride of that wood. It led to a Y-shaped valley filled with bracken, in which couched several enormous Fallow bucks, who came out to feed on the grassy ride in the dusk. They were resplendent in their summer coats and, although little disturbed, were very wild and wary. I would watch them awhile before they saw me and bolted, and then I would return to bed.

I thought of this wood as a permanent bourne; but to go in once, lie low for four days, and go out again is a very different thing from regular coming and going. The keeper would certainly have found me, and probably did not only because it was Bank Holiday week-end. There were other entrances to quieter parts of the wood, but the gates were uncompromisingly locked. Between one and three o'clock in the morning one can generally take a gate off its hinges undisturbed and, after concealing one's traces, stay within until finally discovered and thrown out. One just says that someone must have left it open! However, this, I understand, constitutes breaking and entering, and it was evidently not the sort of thing I could do every evening after work.

Much nearer Oxford lay Wytham Wood, a University property, made famous to naturalists in recent years by the long-term studies of birds and beasts carried out there. It is, I believe, the *locus classicus* for watchers of various kinds of tits. Since I was actually working at the E.G.I. and since I knew David Lack, the then Director, reasonably well, I asked if I could get permission to park in Wytham. I explained that I would not plaster the place with filth – there was a rubbish bin at the E.G.I. where I could deposit my old tins, bottles, and plastic bags. David Lack said he would see what he could do, and I felt hopeful.

However, I had reckoned without the inevitable English desire to be disobliging if at all possible. The answer was 'no', on the grounds (of all the absurdity) that 'if we allowed anyone to do it we'd have to allow everybody'. As if a middle-aged and not undistinguished Civil Servant was to be equated with a tramp! When I want to live peaceably in an area for some time studying, say, Golden Eagles, it is my custom to ask civilly for permission. However, I find this works on only about one occasion in three, though again it is easier in Scotland, where a kilt has many uses.

Anyway, I went and scouted Wytham Wood by day, and spoke to the warden. I found he was an ex-Colonial Servant like myself, who understood precisely what I thought of the situation in Kenya and of my British masters. He could not actually allow me to park in the wood, he said, but he knew many farmers round about and kindly agreed to talk to some of them. In the end, it was all sorted out with the greatest goodwill over foaming tankards in a local pub. I had a splendid parking place under two giant oaks in a hedgerow and went for a walk in Wytham Wood every evening.

Although I had spent hundreds of nights abroad in woodlands and other places where I had no right to be, I had never encountered a Badger. They are not common in Scotland, at least near my home town at Elgin, or near St. Andrews where I was at university. At any event, this was the case before the war, when gamekeepers were twice as large, three times as numerous, and even more destructive than they are now. It was not a matter of climate alone for, after they had been exterminated once, Badgers were reintroduced into the inhospitable environment of north-west Sutherland by a more enlightened landowner, and thrive there to this day. The woods I frequented in eastern Scotland were much more like a Badger's habitat as I understand it; but I never came across one.

Once, in a wood near Elgin, I came on a collie, desperately trying to unearth some animal from beneath the roots of a fallen alder. From the sounds within I guessed it might be a Badger; but though I remained for some time the collie, who was ill equipped for his task, failed in his objective despite encouragement. Neither of us, I am sure, had any right to be there, but he gave me a friendly wag of his tail, as if to acknowledge the presence of another brother of the nocturnal fraternity, and I left him still digging. Nor did I ever see a Badger in that place.

From my camp in the field below Wytham Wood I had to cross a stretch of meadow to reach a small stile from which a barely perceptible path led up through young plantings, full of Fallow Deer, to the ridge top. Here it emerged on an open ride that led on into deep old beechwoods. I worked at my typewriter each day from soon after dawn to breakfast time, then spent the day reading in the E.G.I., and at evening returned and typed again till it was too dark to see. I lived on sandwiches, cornflakes, fruit, cheese, and roast chicken and chips, washed down with cans of beer. I reduced my washing-up to one plate a day and lived well with the minimum of trouble. It cost me, at that time, less than £1 per day including the petrol. Things are not what they used to be!

When the light became too poor for typing I made myself a cup of strong coffee and, after drinking it, set out, rain or fine, across the meadow and up the path, till I reached the ride. Here there was a monument and a seat, erected by some past owner in memory of a loved daughter: a pleasant place to sit and smoke a pipe and contemplate for a while, as the dusk drew in. I used to sit there thinking of my small son so many miles away and wonder how it was with them (well, according to my wife). Then I would knock out my pipe and stroll on down the ride, into the beechwoods, where only a practised night-walker could see what he was doing and move quietly under the canopy of summer leafage.

I guessed there must be Badgers in the wood, but my first meeting with one was a real surprise. I was just inside the deep shade of the beeches, on a dry warm night when the dead leaves and twigs crackled underfoot. I was standing still when I heard quick, running footsteps approaching me. They sounded much like those of a gamekeeper's terrier, and I braced myself for a tiresome argument. I could not see what was coming until they were very close. Then the pied mask of the Badger told me I need not fear any argument. There were two, probably well-grown, cubs of the year. One halted about three metres from me and stared. The other swung broadside-on, fluffing up its grey coat until it looked twice its real size, clearly a defensive reaction. Then they turned and scampered back.

The very next night I met another. Once more, soon after I had entered the beechwood, I heard that quick, light scampering on the leaves. I stood stock still, and this time the Badger, a bigger one, probably an adult female, came right to my feet, stopped, and looked up at me. It was not one and a half metres away as I bent slightly to look it in the eye. It had run along the ride in a confident way as if it had done so every day of its life, to be brought up short by a motionless object right in its path. Though the light was dim under the trees I could clearly make out the pied facial markings and grey back, which doubtless warn other animals confronted with a Badger that this is no creature to trifle with. Indeed, I felt like that myself, for as it looked up at me I half feared it might snap at my ankles, which I guessed would be painful. However, after gazing at me for a good ten seconds it turned and scampered back.

As first meetings with a hitherto unknown animal these were very satisfactory, in a lovely wood on a lovely summer night, all by myself and at close range. They told me, too, that in that sort of dim light a Badger's sight must be very poor, for even the adult had come right to my feet before it even realized I was there, although my human form must have been outlined against the clear summer sky at the entrance to the ride. Either these Badgers were not afraid of man, or they could not see well and were not sure what sort of obstacle they ran away from in the end.

I thought that these Badgers might just have been careless, or inexperienced. But later that night I met another, a much bigger one, probably a boar, near the junction of two tracks in a more open part of the wood. It was working quietly along a wire-netting fence, seeking I know not what, and it too came to within a few metres of me before it saw me. Even then, perhaps because I stood stock still, it did not take fright but, after a pause, went on along the wire. I allowed it to get ahead until it was almost out of sight, then followed silently on practised feet. It was surprisingly visible even in the dark shade beneath the big trees, the grey coat standing out well against the background of dead leaves. I walked behind it for at least a hundred metres. At length, it seemed to realize that it was being followed persistently and drew slowly away from me and the track until I lost sight of it in the gloom. Even then it did not seem to be frightened and may only have moved away on its own purpose.

I did not think that there would be any point in following it further, for I could not really make out what it was doing, even through binoculars, which can in fact be surprisingly useful at night. So I went quietly back along the track and met it again at the junction, or another big one just like it. It was only two hundred metres or so from where I had finally lost sight of it, so I suspect it was the same. Even then, as it came towards me slowly, it did not panic, but just turned and withdrew the way it came.

After that, my nightly walks in Wytham Wood had more of a purpose than merely a little fresh air and gentle exercise after a weary day of paperwork. Every evening I climbed the track to the ride and walked along, hoping to meet Badgers again. Sometimes I saw them briefly, and many times heard that quick, light, terrier-like scampering among dead leaves. However, none ever came as close as the two I saw that first night, nor was I again able to follow one slowly for five minutes or more. Once or twice I tried going out earlier in the hope of finding the sett, which I suspected to be in a valley near where I saw the big boar. I failed; and I could not really spare the time. The reading and the typing, which were my main reason for being there, had to come first in the limited time I had available. I have never worked harder in my life – and I have often worked hard. However, I am still enjoying the fruits of that hard work, and am about to revise the book and, I hope, make it a better and more enduring work than it could be when I first wrote it.

The Badgers could therefore only be a side-line, but I was deeply intrigued by the mysterious beasts, and enjoyed each brief glimpse I had of one. Then, one evening, quite close to the seat, I had another really good view. It was a wild, blowy night with clouds and intermittent spattering rain, and as I walked along the ride rather briskly I heard an animal, presumably a Badger, scuffling in a patch of bracken and brambles. I stopped and, realizing that there was a strong wind blowing from me to it, assumed that it would bolt, for Badgers are supposed to have a keen sense of smell. Not a bit of it. The Badger – for Badger it was – emerged on to the ride within three metres of me; snuffled at something, stood there awhile, and, without so much as a glance at me, turned back into the growth, where I heard its footsteps receding for some time after it disappeared. Apparently, not only did it not see well, but it did not smell well either. I am not going to believe that it could not smell me, for I was only a short distance away and the wind blew straight from my direction. Any other animal likely to be startled by the scent of man would have bolted in haste.

So, either those Badgers of Wytham Wood did not see well or have a keen sense of smell, or else they had been so long unmolested that they had lost any fear of man. It would be nice to think that in England in 1961, and doubtless still today, there were or are Badgers at Wytham that have no fear of man. In the natural state of affairs they need not have any, for they are not the sort of animal that, in the early days before man became the way he is, would have had any reason to fear him.

I saw no more Badgers till 1967, when I was in north-west Sutherland doing a survey of Golden Eagles. This was the area in which they had been exterminated and then reintroduced, a wild land of rough mountains, then naked of trees except for small birch shaws nestling in protected valleys. I was *persona grata* in that part of the world, where the first keeper I spoke to said, 'Ou, ay, ye can use the bothy at the head o' the loch'; and, when asked if there were any small lochs where I might catch a trout for my breakfast, replied, 'Man, if ye can walk three hunner yairds ye can fish where ye like.' The only exception to this general rule was punished to the extent of a salmon and a good sea trout, though I was too old to allow myself the luxury of rousing him from his bed in the middle of the night and making him run about the bogs to catch me.

It was my custom, for some while, to eat my dinner in a small cave beside the Laxford River, one of the few places where I could get out of the wind, which tended to blow out my small gas stove in the open. The bottom of the cave was covered with the empty shells of freshwater mussels, left there by the tinkers who annually visit the Laxford to search for 'Spey pearls', as they are called in eastern Scotland. Apart from the midges, which got under my kilt by the million, despite repellents, it was a nice little holt. I carried my dinner, stove, pans, and cutlery there in a haversack, and after eating, carried them back to my Peugeot estate car in the back of which I was living. It was not a good Snail, but it sufficed.

One such evening, walking back quietly along the road, I met a huge Badger, the biggest I have ever seen, probably an old boar. I followed him, ten metres away, until he turned off the road and went up a rocky hillside, apparently completely unconcerned and unaware of my presence. Another time, when I was sitting in the back of the car writing up notes, this huge boar, or another just like him, came into the little quarry where I was parked and snuffled about among refuse left by tinkers within a metre or so of me. These Badgers were not much afraid of a quiet walking man or a parked vehicle either, yet they were living in an area where Badgers had once been persecuted to extinction by the relentless gamekeeping fraternity of the nineteenth century. The head stalker of that part, Willie Scobie, said later that he had nothing against Badgers – they did not eat deer or catch salmon, and if they did eat a few grouse eggs there were not that number of grouse to bother about anyhow. But then, he was an enlightened man for a gamekeeper, and his employers did not force him to be otherwise.

I have only seen one Badger from that day to this, poking its nose briefly out of a sett on Chinnor Hill. However, I have read something about Badgers, and I spent several evenings watching the BBC programme *Badger Watch* at a sett in Gloucestershire. From my reading and watching the television I learn that a Badger has a keen sense of smell and that if it scents a man, it will bolt at once. Also that one must ascertain what Badgers do by piecing together the traces they leave behind them on their nocturnal wanderings, for it is impossible to follow one without frightening it.

All I can say is that, although I cannot pretend to know very much about Badgers, the only Badgers I have seen, in two different parts of Britain, did allow me to follow them quietly: one on such a light night in the north of Scotland that I could clearly see what it did; and one, at least, that did not have a keen sense of smell, or did not immediately bolt when I was in easy smellshot of him. Perhaps they were different kinds of Badgers, or perhaps I am not an ordinary man. Certainly, as I am now quite old enough to admit, I am better than most at skulking quietly about in woods where I have no right to be at all.

D. TISCHLER

3. Beavers

A great deal of time is devoted nowadays to the study of beavers. They have appeared on television and in films by Walt Disney, in the pages of the *National Geographic Magazine* and in many books. From all these sources one may gather that they are nice creatures that do no one any real harm and in many areas do a lot of good. It is just bad luck for them that they have a skin with usable fur, as it is for the leopard, the Grevy's Zebra, and many species of otter. Because someone wants to wear their fur, excuses of one sort and another are devised to kill them.

There are actually two species of true beavers, the European *Castor fiber*, and the American *Castor canadensis*. The European Beaver may have thrived at one time over a very wide area, but has been exterminated in much of the western parts of its range, and survives in some curiously isolated colonies, such as one in the lower Rhone Valley. Its main haunts are in the great expanse of northern woods, extending from Scandinavia to the Urals (and no doubt further east), though even here it was exterminated in Sweden, but survived in Norway. It is also common in the wooded steppes of the Soviet Union round Voronezh, and is there exploited for its fur. Beavers' pelts are valuable. An animal spending much of its time in cold water, such as an otter or a beaver, can be expected to produce a nice warm coat. At one time there may have been some good excuse to hunt beavers on this account; but nowadays synthetic fibres have really made all such furs unnecessary, at least for female adornment, though they may still have uses for special purposes.

The American Beaver has fared much better, no doubt because North America has not been inhabited by destructive European man for quite so long. It was really only in the nineteenth century that most of the American West was colonized and inhabited, at least by large numbers of men, and the enormous wastes of the boreal forest in Canada could stand limited exploitation by trappers for quite a long time. There are still people in North America who make a partial living by trapping Beavers; and excuses are still found for what is nowadays an inexcusable and cruel practice. However, Beaver-trapping is not so prevalent in America as it was; and Beavers are still very abundant in many areas. The American Beaver has a namesake, which is not a close relative, the Sewellel or Mountain Beaver; it just looks a bit like a beaver and is also aquatic. One of the most primitive of rodents, it lives in mountain streams near the Pacific Coast up to 9,000 feet (2,700 metres) above sea-level. It burrows, and stores food underground for the winter, but there its likeness to the true beavers ceases. It is the sole member of its rather primitive family, the Aplodontidae, while the true beavers form another family, the Castoridae. All are rodents, the true beavers among the largest existing.

As most people now know, beavers build dams of logs across small or even quite large flowing streams, so impounding small lakes. Towards the centre of these they build quite elaborate houses or lodges, where they live relatively dry and warm although the entrance is under water and the beaver can only reach it by swimming. Sometimes they live in burrows in river banks, generally where their population is rather thin. Moreover, they are provident creatures and cut and lay by a supply of food sufficient for the winter which, in the areas they inhabit, is long, harsh, and cold, starting in October and continuing till April. At this chore, and at cutting and carting logs for the dams they build, they work very hard – hence the phrase 'working like a beaver'. They do not always work, but in season must work very hard or die.

Like the Antbear or Aardvark, the beaver's activities, designed for its own benefit, have tremendous beneficial side-effects for many other animals and birds. Their dams often impound a considerable sheet of water, or actually raise the level of a larger lake. They are usually placed in gently sloping valleys where the speed of streams draining from mountains is not too great. In calm water the beaver dam runs straight across the stream, but in strong currents is bowed upstream to withstand greater pressure. Evidently beavers have an inborn capacity to understand stresses and strains, far better than that of their chief persecutor, man. The sheets of water so impounded may be up to five hundred metres long, but are usually smaller; those I have seen have varied from fifty to two hundred metres long. The dams themselves may be up to three metres (twelve feet) high, usually lower, built of wood, where necessary weighted down with heavy stones. Mortared with clay and dead leaves, the downstream side of the dam is at an angle, the upstream nearly vertical. They could hardly do it better if they had been taught, considering they have no tools; and the sheets of water permit many other creatures to live where otherwise they could not.

Beavers are also among the original conservationists. In northern latitudes erosion of mountains takes place largely through the fragmentation of rocks in winter frosts, and the violent run-off caused by snowmelt in spring. Where, as in Scotland, this run-off comes from grass or heathery slopes it carries very little soil. The streams run high but clear, though spring spates do undercut banks and gradually carry them away so that, in summer, quite a small river like the Feshie runs in a wide torrent-bed of stones. When snow-melt comes from bare slopes of earth at higher altitudes it has much greater erosive power. The soil, which becomes sodden anyway, is inclined to slide and run off, and such mountain streams then carry far more silt.

Where beavers have been working this soil and silt is carried into their dams and there caught and held. Over centuries the silt builds up and fills the dam. Beavers then build higher so as to ensure they have enough water for their own ends. More and more silt is trapped. Luxuriant beds of water-weeds develop on the rich muddy bottom of the dam and help to hold the silt. Finally, a metre or so of soil may have built up; a swamp develops in the once open waters; and the foundation for a future fertile alpine meadow is laid.

Of course, it takes a lot of beavers a long time to do all this, but essentially they employ the same principles as would be recommended by any latter-day soil and water conservation engineer. The basic effect of the dam is to slow down the run-off water. Since the amount of sediment and the size of the particles that can be moved by water depend on the square of its velocity, slowing down the stream means that first sand and gravel and later fine silt are deposited instead of being carried downstream to waste. Thus it is carried to waste in countries not inhabited by beavers or conserved by soil conservation engineers. In really slow-moving water, such as that found in any beaver dam, even fine silt is filtered out by the weed-growth on the bottom, and the water becomes quite clear.

Beavers cannot do their beneficial work everywhere. They must have suitable softwood trees with which to build their dams, and there are many situations which would defeat even their remarkable technical abilities and their enthusiasm for hard work. The trees they need for their winter food supply and for dams only grow at certain altitudes. I can think of many valleys in Scotland ideal for beavers if only there were trees; perhaps there were once, but they have since been destroyed by men and sheep. In other places, such as a rocky gorge with leaping waterfalls, beavers could not control the violence of the stream even with the necessary trees. However, there are plenty of places in Europe today with the right trees and the right slope but no beavers – largely because they have been exterminated by man. I believe that they are now increasing in such countries as Sweden.

If beavers did their good work in Europe long-ago, they were later exterminated, and as a result their works collapsed. The result would have greatly accelerated run-off in snow-melt, for the dams need continual maintenance if they are to keep holding up water. If the beavers do not maintain them, no one else will. An abandoned dam collapses; and a sudden torrent of mud, bits of log, sand, and weeds rushes down into the next dam, overstrains that, and down it goes too. Later in the year there is nothing to hold up the water and the weeds that once grew on the bottom of the dam shrivel and die, leaving bare, loose earth. More of that erodes in heavy rain. Wiping out the beavers in a single valley could result in a catastrophic series of floods carrying away all their works in a few days, or at most a few years. Once gone, the soil cannot be replaced from where it came, on the mountainsides higher up.

The work of millenniums by beavers can be destroyed by a few fools – always human – in a few years. The human race is generally incapable of observing simple facts of life that affect its own salvation and survival and it will pay bitterly and deservedly for its stupidity in these matters before I am dead. It is paying now, in many countries. Humans like blaming catastrophes such as famines or floods on God; but all too often it is their own fault. Occasionally an earthquake or a volcano can properly invoke the insurance companies' phrase 'Act of God', but too often catastrophes are due to human stupidity and lack of observation and, nowadays, a desire to be popular and 'influence people'. After we have gone, some beavers may still be around to repair the damage bit by bit. But it will take them a long long time, and in many countries they do not exist, and never will.

Beavers are not easily seen in Europe, but there are still plenty in North America. I first went to the States in 1958, on a U.S. AID grant to study Range Management. Since three-quarters of Kenya is composed of semi-arid grasslands and semi-desert I wanted to learn about this science from the fountainhead, which is in the U.S.A. Since there is no semi-arid rangeland in Britain, British Colonial officials at that time knew very little about Range Management. Nowhere in my training did anyone ever tell me anything about it. However, out of 582,750 square kilometres (225,000 square miles) of Kenya, about 466,000 (180,000) is rangeland, inhabited by livestock, wild animals, or both. About two-thirds of all the livestock, nearly all the celebrated wild life, and a sizeable segment of the human population live in this rangeland. In the late fifties it was largely controlled by veterinarians, whose single-minded objective was to control stock disease – never mind whether the animals died of starvation later, as they often did. I felt that this was a situation that could and should be improved upon, but it seemed clear that no veterinarian wedded to his needle could or would change it. So I asked for and got a small grant to go to the States for about six weeks in the summer.

I went straight from Devon to Washington, and was met at the airport by a well-meaning woman whose face fell when she saw I was not black. I was supposed to attend a four-day orientation course to learn about the U.S. constitution and Abraham Lincoln, but I cut that and spent my time in the art gallery and in the Smithsonian Museum. I took with me a unique film I had made of flamingos for BBC television for my first 'Look' programme to show to the *National Geographic* magazine. They could not be bothered to sit through it, though they certainly had seen nothing like it before. Generally, however, I had a good time in Washington, but I was glad to get out of it, for in late July it was stinking hot and humid, and my grant did not run to an air-conditioned hotel.

An itinerary of more or less continuous travel had been arranged for me, through Oklahoma, Texas, Arizona, California, and Utah to Colorado. I hoped I might be able to snatch a few days here and there to look at the Grand Canyon and the giant sequoias, but I had reckoned without the vast distances and shortage of funds. The only famous place I managed to see was the Bear River Refuge in Utah. But I fell in love with the U.S.A. and its people, all of whom, without exception, were kind and hospitable to me. I think they found it a bit odd that someone of my standing (I was then Deputy Director of Agriculture) should have come to learn like a student, and they went out of their way accordingly. Subsequent visits, however, have not altered, but only improved upon, first impressions, and now I have as many friends in the States as anywhere. So, Uncle Sam, here is one satisfied customer!

My grant enabled me to 'just about get by', as they told me, and it was stretched by the kindness of many of the people I met along the road. I travelled at week-ends by train and Greyhound bus; and the most horrible place I have ever seen in my life, excepting the mudflats of Lake Natron, was the main Greyhound bus station in Los Angeles. Here I had to get a midnight bus because my train – the celebrated Sunset Limited – was seven hours late, and I missed my connection. Trains in the U.S.A. mostly ran east to west in those days. I made one fearful north–south journey from Woodward, Oklahoma, to Sonora, Texas, an experience I would not willingly repeat. I got out at a place in Texas called Lubbock, in dead flat semi-arid country, to have breakfast. I bought a paper sarcastically named the *Lubbock Morning Avalanche* from a man in the square. I observed that it was a dead-and-alive place, to which he replied morosely 'Gee, you oughta see it when there's a sandstorm blowin'.' My friend Emil Urban tells me that Lubbock is not as bad as I thought. Most of it was destroyed by a tornado some years ago, but I should still approach it with some reserve.

I had brought my trout rod from Devon, hoping to find some fishing along the way. It was a two-piece Hardy split cane, given to me by a girl-friend for my twenty-first birthday, and cherished ever since. I had been using it with great success to catch sea trout in Devon just before I left. A redcap took it from me at Washington airport, and I, not knowing American customs, let him. He flung it carelessly on a conveyor belt, which had a right-angle bend. I was too late to save the rod when it reached this bend, and though the top was flexible enough the butt snapped. Frantic with fury I took it to the airline desk. 'What's it worth?' they asked, and like a fool I said, 'Twenty-five dollars.' They would have gladly paid me fifty to get rid of me. I snatched the rod, cursed the redcap fluently, and, when I had calmed down, took it to a Washington tackle shop. They made a very adequate repair for ten dollars; and many's the good fish I have caught with it since. It is all that remains of my twenty-first birthday presents, apart from a set of dress studs that I shall never use again. I no longer own the necessary suit.

I carried the rod all over the States with me but had no chance at all of using it until I reached Utah. I landed up in Logan on Labour Day week-end, with four whole days to kill in a hotel where, in those times, one could not buy alcoholic drink or even coffee. I went to the Bear River Refuge one day, but it was not at its best between the breeding season and the influx of autumn migrants.

A lovely trout stream flowed out of the Logan canyon, and in its gin-clear waters I saw fat trout idling. So I asked if it would be practicable to take out a trout licence and have a day's fishing. It was possible to buy a licence, but they told me there would be a fisherman for every ten metres of river, trying to catch the fish then being put into it by the thousand. Instead I elected to go for a walk in the Wasatch Mountains. This in itself astonished everyone that I passed on the road, and at least six people wanted to give me a lift. However, evading them I climbed to a high ridge, where I found a Golden Eagle's nest and could look down into the canyon. Sure enough, where the road ran alongside the river there was a car parked and a fisherman casting, every ten metres. But at one point the river ran round a rocky bluff, while the road cut across it. On that short stretch of river no one was fishing. I could have fished there all day and probably taken a good bag, because I doubt if those relatively inaccessible fish ever saw a fly. I should have had to scramble to reach several pools.

I had worked hard, learned a lot – including how not to run a Range Management Department – and had a good time. But I had so far had no free time at all. My last stop was at Steamboat Springs, Colorado, where I stayed with Tom Eaman and his wife. Here I earned a welcome extra fifty dollars giving a lecture to the pupils of a superior private college. It was news to them, and apparently incomprehensible, that there were pastoral nomads in north Kenya at that time who had absolutely no use for money as such. My time at Steamboat Springs was the most relaxed and pleasant I had had on the whole trip. For the first time it was cool, and I had a little spare time and a chance to see something of the Colorado Mountains. I have had a soft spot for Colorado ever since.

There came a Saturday afternoon with a whole six free hours before I caught my train. We all went up into the mountains to a dude ranch where the proprietor ran pack-horse trips into the wilds in the autumn. He saw I was no dude and hoped I would be able to make a real rugged trip with him some day. The others wanted to ride, but I hate horses as I always fall off them and hurt myself badly. I was very content to take it easy and to fish in what they said was a fair trout stream. The rod would have an outing at the very last.

When I reached the stream I was disappointed, for here was no rushing clear water like the Logan River, but a muddy, almost stagnant ditch overgrown with willows in which one could hardly cast at all. It was no more than about a metre wide and obstructed everywhere. I like to be able to use my rod freely, on a big river, casting long and straight, so this was not really the place for me. However, the afternoon was so lovely that I was content to walk up the side of the brook, casting in a desultory way when I found a clear stretch. I caught nothing.

Not that I was bored. Far from it, in fact. It was a glorious crisp sunny afternoon in September, and the first frosts had bitten the aspens so that the clear sparkling gold of their quivering leaves stood out against the dark backcloth of pines and firs. It was hot and nearly windless at four o'clock, but I could sense the coming chill at sundown. There were jays and squirrels, storing seed for winter, and an occasional mule deer to be seen. Like most fishermen I was quite happy to be idling along in a lovely place without catching any fish. As the sun declined, the gold of the aspens intensified till they literally blazed. It was my first taste of the glorious American autumn colours. We have good colours in Scotland too and the same clear mountain air, but too often a wild wet autumn soon blows off the leaves and makes everything soggy underfoot. I have seldom been in a more beautiful place, and my conscience was clear, for I had done what I came to do.

Then I came to a big Beaver dam and saw at once why the water lower down was so muddy. They were there, working with their traditional, almost frenzied energy to store away food for the winter. Their comings and goings had stirred the bottom mud to foul the stream as it ran out of the dam. They were fetching their supplies from a grove of nearby aspens, and in a few minutes I had found trees recently felled by their chisel teeth, logs cut to size ready for transport, green branches and twigs, and had seen my first Beavers in the dam. The alarm signal of a beaver is to slap the flattened tail against the water surface, and a couple of these sharp slaps gave them away. I could see their heads regarding me from near their lodge, in the central lower part of the dam, in water about a metre deep. For the time being they stopped working and disappeared.

The edges of the dam were grown with small willows, making casting almost impossible. However, where the stream ran in at the head of the dam the water looked clear and there was an open patch, more or less free of reeds and rushes. I went there and found I could cast a long straight line down the length of the inflow and draw the fly slowly backwards between beds of weed. Each cast occupied a couple of minutes, and as I stood there, casting steadily, the shadows began to lengthen and I merged into the landscape. The Beavers came out and went on with their work.

They swam about the dam, carrying sticks held in their teeth, and though they did occasionally notice the movement of my rod as I cast, my subsequent stillness reassured them. They grew more and more used to me as they went to and fro, and finally ignored me completely as the dusk deepened in the long calm evening. It was wonderfully peaceful, and I was able to watch them doing their celebrated work right in front of me for about an hour.

I caught nothing for a while, but then, as I was drawing the fly back close to my feet, a good-sized Rainbow Trout took it, and I kept him successfully out of the weeds and guided him into the bank. This was followed in quick succession by a half-pound Brown Trout (225 grammes) and then by a Brook Trout, a species I had never caught or seen before. Two more Rainbows completed the bag, the best about a pound (450 grammes), before the fish stopped taking. The Beavers had never stopped working all the time I was there, and I could scarcely see their heads against the dark steely water when I left. There was a chorus of slaps as I finally disappeared, for by then I was cold and glad to move swiftly away. In that hour I had seen most of the things I had read about regarding beavers at that time of the year, and they had scarcely been disturbed by me.

I walked back briskly in the last of the light to rejoin my friends. The brilliant gold of the aspens had dimmed and a purplish haze hung over the valley. I could feel the chill that presaged a sharp frost in an hour or less. Our rendezvous was one of those square concrete tables with a ring of concrete benches round a fireplace in which blazed a huge pile of logs, burning down to coals for a 'cook-out'. On the table stood a huge square bottle of bourbon, containing, I believe, a U.S. gallon, and on a grill, ready for the fire, was a double row of succulent-looking Elk-meat hamburgers. They all marvelled that I had caught five good trout in a place where hardly anyone ever caught anything – as they now admitted. I am not a polished or artistic fisherman, but I generally catch fish, and I felt that the honour of Scotland and Kenya had been upheld.

I was sharp-set after my evening's fishing, but without the bottle of bourbon those Elk-meat hamburgers would have been virtually inedible. Anything resembling more closely a ground-up football I never ate. The others professed them marvellous, but, like the dish of tea-leaves offered to the Emperor Ming Wang by the discoverer of tea in the Kai Lung story, they had no perceptible flavour and produced a feeling akin to suffocation. However, hunger is the best sauce, and the fire and the bourbon combined to produce a heady glow of well-being. As the level of the bottle dropped we sat round the fire replete, tongues wagging as we yarned of this and that. Good fellowship and happiness prevailed.

I have never seen another beaver since, though I have seen many of their dams in the Adirondacks, Colorado, and elsewhere. But never shall I forget that golden Colorado evening, when I could watch them going about their autumn work unafraid of me, while I indulged one of my favourite pursuits with unexpected success, basically just because they were there and had built that nice dam for me to fish in

4. Chimpanzees

Chimpanzees are now also among the most intensively studied of all mammals, partly because they are man's nearest living relative. Many people know how Jane Goodall and her various associates have been studying Chimpanzees at the Gombe Stream National Park in Tanzania for many years. From modest beginnings as a one-woman exercise this has now built up into a major research station with, no doubt, a long waiting-list of applicants to go and work there, and results analysed by computer. There seems to be a longing among many young people to take part in other people's research from an organized base instead of starting on their own to gather facts ready to hand. I was once asked in a hushed tone (by an American) whether I had ever 'understudied' Jane Goodall or Richard Leakey, and I had a certain satisfaction in replying that I had so much independent research of my own on the go that I saw no reason to understudy anybody. Indeed, I would not be good at it.

My own, one and only, encounter with Chimpanzees occurred in 1964, when work at the Gombe Stream was beginning to intensify and become known. There was no point in my attempting any research on Chimpanzees in detail, even if I had had the particular desire to do so, for there were others, well funded and in a much better position to do it than I. There is plenty to do in Africa without falling over each other's feet. However, I was gathering material for my book on Africa in the series called 'The Continents we Live On', published by Chanticleer Press. When I was stationed in West Africa I had never seen a Chimpanzee, since I was never posted to an area where they occurred. However, I had read Du Chaillu and others on the subject of the great African apes and I wanted to see a Chimpanzee for myself.

The nearest place to do so was in the Budongo Forest of Uganda, where some preliminary research work had been done on them at that time. Scientists feared that the Forest Department's policy of eliminating 'useless' tree species such as parasitic figs by the use of herbicides, so allowing (in theory at any rate) more 'useful' trees to grow, would threaten the Budongo Chimps. Even before independence forest reserves were often under pressure from British administrative officers, who ought to have known better. As a result the Forest Department sometimes tried to justify continued reservation of forests by developments that would produce more valuable timber products and be justifiably economic. Few people seemed to think that a tropical forest was a thing that ought to be preserved just because it was a forest. While such an 'enrichment' policy could probably produce more boards for furniture, it could be bad luck for the Chimps, who happened to like the fruits of some of the 'useless' trees.

So I spent two days in the Budongo Forest, looking for Chimps where the rangers said they were regularly seen. I saw none, nor even any recognizable traces of them. From there I was going to Fort Portal and to the Bwamba Forest, but *en route* to Fort Portal a piston of my Land Rover's engine cracked and I limped into Fort Portal making a horrible din and wondering whether I should be able to afford the repair. I stayed three days in a hotel with nothing to do, and passed the time by writing in longhand much of another book, *Ethiopian Episode*.

Then I went on into Bwamba and to the Semliki Valley, where I finished the book under a nice big tree close to a road quarry, which I had found by the simple process of disregarding a 'No Entry' sign. Even in Africa it is odds-on that anything saying 'No Entry' is worth investigating. In this case it was kindly meant, to help travellers avoid a track that only led a few hundred metres. However, I was able to finish my book under that nice tree, with large herds of buffalo, elephant, water-buck, and Uganda Kob all round me, and lions roaring each night as I sat by my fire. It was quite idyllic, an old-time Pleistocene experience.

In the Bwamba Forest I found signs of Chimps, but saw none. There were nests in the trees that could only have been made by them. I took some photographs of forest plants and scenery, hoping they would be published, including what should have been a splendid one of two Colobus Monkeys sitting in the sun on the end of a *Terminalia* branch, eating the bursting buds, silhouetted against a misty early morning in the valley below. It told the whole story of Colobus Monkeys, which are among the most delightful of primates until you get within range of their belch. But I saw no Chimps and thought I would have to return to Kenya without seeing them.

I had intended visiting another western forest, the Bugoma Forest, on my way to Fort Portal, but had been denied this by the fault in the Land Rover. Now, having been able to pay the remarkably small bill, and with a rejuvenated vehicle, I decided I would have one last try, on my way back to Kampala and Kenya, to see a Chimp in the Bugoma Forest. I could only spare part of one day, for I was under pressure to finish *Africa* and about to set off on a long journey to Cape Town and back in the Land Rover, but I thought it was worth a try.

I drove in by the entrance track at about ten in the morning. I had absolutely no idea where to look, and thought it was a pretty forlorn hope anyway. Fortuitously I turned right along a fork of the track skirting the edge of a forested valley in open savanna, and in the top of a huge fig-tree, about 400 metres (a quarter of a mile) inside the forest, I saw black bodies moving about, and heard unmistakable high-pitched yells and screams. The fig-tree stood well above the rest of the forest growth, and my Land Rover was in clear view from it. So I did not stop, thinking it just possible that the Chimpanzees – for such they undoubtedly were – might pay no attention to a vehicle that kept on going, though they might be alarmed if I stopped and looked at them with binoculars. I drove on, and after about one and a half kilometres the track entered the forested valley. Here, in a place of deep quiet shade, I parked near a small stream, left the Land Rover, closing the doors as silently as I could, and began to stalk the Chimps.

I had no difficulty in locating the fig-tree by the chorus of cries of all sorts which were clearly audible once I was within 800 metres (half a mile) of it. However, I knew that all primates have very good vision and that to stalk a big group of Chimpanzees without being seen would require all the skill I had developed in past years when walking the midnight woods on more nefarious pursuits. As I approached the tree, I crept forward pace by pace, keeping under thick leafy cover where I could, on the principle that if I could see a Chimp it could also see me, though it certainly could not hear me above the din the troop were making. I kept my tell-tale white face pointed to the ground, beneath the brim of my hat, and cautiously peeped through chinks now and again. At length I gained a group of small trees overgrown with creepers, right under the canopy of the fig-tree, and through small gaps in the foliage found I could watch the Chimps fairly well, though they evidently had no idea I was there.

The fig-tree was one of the biggest I have seen. It must have spanned quite sixty-five metres (seventy yards) between the tips of its branches, and it rose like a huge umbrella of leafage out of the rather stunted forest in that part of the valley. Eighteen metres (twenty yards) from me was its bole, a good four and a half metres (fifteen feet) thick, buttressed at the base, and rising in a symmetrical golden column to a main fork about twelve metres (forty feet) above ground. There it spread first into two, then into more huge, golden-barked limbs which arched out and over the forest undergrowth, beneath a dome of foliage, thin now in the dry season, letting in the sunlight. These branches were covered with trusses of big ripe figs, and the Chimps were feeding on the fruit high in the canopy. Since even the lower limbs of the fig-tree were above the understory forest trees, I reckoned they could only have attained the branches of the fig by shinning up a vertical forest tree that rose from the leaf-mould about twenty-seven metres (thirty yards) from me and almost touched one of the large arching limbs about fifteen metres (fifty feet) above ground. Even the wide span of a big Chimp's arms would not have encircled the main bole of the fig, which was one of those species with scaly slippery bark, very difficult for a human being to climb.

I do not know how many Chimps there were in the tree, for I could not count them, but I should say not less than thirty and perhaps as many as fifty. I know now that they must have represented many family groups, that an expert on Chimps could have recognized them as individuals, and that someone interested in particular aspects of behaviour, say the relative frequency of scratching by males and females, could have got a nice lot of data which he could have put into a computer to come up with a statistically significant figure. Mercifully, I was not concerned with that sort of science, nowadays the fashion. To me they were just Chimps, and marvellous in their own right, never mind their 'interactions'.

All sizes and sexes were feeding greedily, squatting on the big limbs of the fig. They picked the fruit off the bunches with both hands, and stuffed them hand over hand into capacious lips and cheek pouches. Their eyes were apparently bigger than their mouths, and probably their bellies, for although their dark and flexible faces were visibly crammed with juicy figs they strove to thrust in more even as they chewed. A sort of rain of bits of fig spattered on the leaves of my enveloping creepers and sometimes on me, and the rootlings of bush pig in the leaf litter showed that they came there at night to enjoy this feast of remnants. Many figs would have fallen anyhow; the Chimps just speeded it up a bit. I have never had the pleasure of witnessing a pie-eating contest, but I never saw a more abandoned example of sheer unbridled gluttony in my life. I guessed the Chimps must have been feeding there for days, for there were dried fragments of chewed fig all over the forest floor and sticking to every leaf, and the undergrowth was flattened as if they had fed on the ground too.

It was a comical, and yet a splendid, spectacle. No zoo Chimp I ever saw looked like these magnificent glossy black-haired apes in their natural surroundings. Most appeared to be females and adolescents, but there were at least four big males. These kept rather apart from the yelling horde of females and younger animals, ate more sedately, and moved more slowly along the limbs. Others gave way to them; but when they reached a particularly juicy lot of figs these big males crammed them in just like the others, cheeks bulging, juice dripping down, and faces plastered with orange goo just like any youngster. Why they ate quite so greedily and made such a mess of themselves I do not know. They would have swallowed just as much, and got just as many good figs, if they had been a bit more choosy and selective. It was as if they had little time and were trying their hardest to cram as much in as possible before they were caught. Yet, in that spot, it appeared to me that they had all the time in the world. It was a fascinating spectacle anyhow.

There is a limit to the amount of food that can be crammed into a primate stomach at any one time, and after a quarter of an hour or so some of the Chimps had evidently had enough. They sat and rested on the great golden sunlit boughs, and some shuffled up to their parents or elders and sat beside them. One enormous Chimp, who doubtless was a high-ranking male if not the dominant of this group, came half-way down the tree and sat at his ease in one of the main lower forks. His legs dangled loosely below, his belly bulged, his face was plastered as with lemon curd, his mouth a wide orange crease. Yet he was a magnificent sight as he sat there, steadying himself by stretching out an arm on either side to touch with clenched knuckles the widening fork. He was no more than thirty metres from me, but had not seen me, and in fact was quite unsuspecting of any alien presence, more concerned with the doings of his friends and relatives above than with anything on the ground.

I had several times cautiously and slowly pointed my camera with its long 300-mm lens at feeding Chimps through small gaps in the canopy, but had concluded that any photograph I could take would be just a mass of undecipherable soot and whitewash against the brassy sky. But what a picture that big Chimp would be! I could not stay much longer anyway, for I had to be well on my way towards home by nightfall. It seemed that activity was dying down and that the Chimps would take a rest, so that I had seen most of what I was going to see until later in the day. That photograph of a great wild Chimp at ease in a marvellous setting would be better than anything I had ever seen published. It was certainly worth a try!

I made ready, rose slowly to my feet, and crept forward to what seemed a strategic opening in the leaves. I did not fail to reckon with the sharp, full-colour binocular vision of the Chimp, but I thought I might take one, perhaps two, quick pictures before his eyes fell on me. He was too alert for me. Although he was looking up at his companions most of the time, his eyes also flicked groundwards now and again, and before I could crystallize the focus he saw me. An expression of horror and alarm came over his plastered visage. 'Wraaaaaa,' he yelled; and again, 'Wraaaaaa.' He could not have said more plainly ''Strewth' if he had tried. He leapt to his feet and led the exodus.

The Chimps, as I thought, had gained the fig-tree by climbing the vertical forest tree near one of the limbs. Now they all elected to come down the same way. I could not understand why some of them did not just run out along a big limb and cast themselves into other leafy branches below, as they should have been able to do.

A moment or two after the big male had yelled, panic had seized the whole troop, and every limb of the fig became a sort of conveyor belt of hurrying black bodies. They had first to descend into one of the main forks by various routes and then run out along the arching limb of the fig to the escape tree. They got in each other's way, and some even slipped, but none fell. The chorus of high-pitched yells and shrieks was deafening, and a rain of figs from the shaken limbs came down to the forest floor.

Reaching the trunk, which I too had now approached more closely, the Chimps came down it at speed and, landing on the ground, ran rapidly away on all fours into the undergrowth. The smaller and less expert among them, or mothers burdened with young, came down swiftly, hand over hand, casting anxious glances at me over their shoulders. The real experts, or tigers, slid straight down the trunk, barely grasping it with all four hands and feet, just as firemen hastening to reach their machine slide down a polished steel pole. They took only a second or so to descend the fifteen metres (fifty feet) of smooth branchless trunk in one superbly controlled agile movement, landing with a heavy thump, and at once taking to their heels like the rest.

In less than two minutes from that first panic-stricken yell the great fig was empty of Chimps and the forest fell silent except for the barking of an Olive Baboon some distance away who had heard the commotion and thought he ought to give tongue, though he had no idea at what. A few minutes later things had returned completely to normal. The plantain eaters, glossy starlings, and barbets that had been sharing the feast with the Chimps had returned and were feeding again unconcerned. The Chimps had made themselves scarce, and I heard them no more.

I could now take time to look around on the ground and observe the signs, which is always worth doing after such an encounter. I walked all round the tree and found several well-trodden paths leading from the vertical access trunk. I concluded that the Chimps must have been feeding there for some time and that they always approached this prolific fig by the same route, for there was no evidence at all that they had climbed the main bole, or any other tree. No doubt the useful access tree had been well known to the older among them for many years and used year after year for the same purpose. The upper surface of the limbs of the tree must have been polished and slippery with the passage of hundreds of skilled grasping feet and hands, but I could not shin up to see.

The understory trees contained numerous nests, suggesting that at least some of the Chimps had fed late at times, and roosted close to the source of food. It struck me that a Leopard – and there were at that time plenty in the Bugoma Forest – could very easily have caught a small Chimp on the ground as it approached along one of those footpaths. Yet I saw no sign that the Chimps approached what was evidently a favourite food supply, perhaps for several weeks each year when the figs were ripe, by any arboreal route. It struck me that they took an unnecessary risk rendering them liable to predation by this cunning carnivorous cat.

The much bigger Gorilla apparently regards Leopards with indifference, but even an adult male Chimpanzee, weighing maybe 65 kilogrammes (143 pounds), should be within the powers of a ferociously armed adult male Leopard, weighing about the same, and frighteningly strong for his weight. A young Chimp, straggling away from others or playful, should be easy meat. In rather superficial reading on the subject I have found little about the inter-relations between Chimps and Leopards; but perhaps if a Leopard tries to kill a young one the big males come to the rescue, as will the much smaller and less powerful baboon. Adult male baboons join together to repel a common enemy – as I found to my cost when my bull-terriers pursued them. They can present a formidable united front against a Leopard which can cope with one, but not six at a time. What happens when, or if, a large Leopard tries to kill a young Chimpanzee? Does the nearest big male come to the rescue, as you or I might? I should like to know.

Reading a little about Chimpanzees later on, and especially about Jane Goodall's early and frustrating attempts to get even a fair view of Chimps feeding for only a few minutes in good light, I realized what a supreme piece of luck I had had. To see them as I did, from directly below and at such close range, of all ages and sizes, and all guzzling with comical abandon on superabundant ripe fruit, unaware of my presence, was one of the greatest short hours of my life. I have never seen another Chimp from that day to this, except in a zoo. Yet I can instantly conjure up the vision of that magnificent male seated at ease in the fork, steadying himself on either hand with a long black-haired arm, and his look of horror as he detected me with my poised camera before he bolted.

Maybe I missed the picture of a lifetime; but if I did, I would not have got it back anyhow. That particular film, containing also point-blank head-on photographs of a bull White Rhinoceros and the marvellous dawn shot of the Colobus on the top of the *Terminalia* bough was misdirected by the firm of developers. I received instead some unwanted pictures of people playing golf at Aden, and of a voyage to Australia on the liner *Canberra*, with simpering females in deck-chairs beside the swimming-pool. Why does it *always* happen to the best photographs one takes? At least I did not lose a never-to-be-repeated shot of that Chimp, and I can always picture his magnificence in my mind's eye.

Peregrine Falcon
Ayres' Hawk Eagle
African Crowned Eagle
feeding young
D. TISCHLER

5. Eagle Hill

Eagles were not my first love among birds. They were water-birds, waders, and the spectacular colonial seabirds of northern Scotland. Nowadays, if for some reason I am deprived of eagles, I turn back to water-birds or seabirds again. Indeed, I should be torn between Great White Pelicans and Tawny Eagles, if I had to choose. However, in 1937 I saw my first Scottish Golden Eagle, in the Corrie Fee of Glen Doll in Angus, and a few minutes later had found my first eyrie. Eagle fever instantly bit deep, bit hard, and has not left me yet. Although the beauty of flamingos and the weird ways of pelicans may divert me from time to time, it is eagles that have been my major interest for forty years, eagles for which I am best known as an ornithologist, and eagles which still delight me as no other bird can.

I wrote long ago: 'The eagle threads the paths of the upper air in a manner eclipsed by no other bird.' No other bird so perfectly combines grandeur and power with surpassing grace of flight. To me, nothing can really compete with the sight of a Golden Eagle soaring in blue sky above snowy cornices, or the piercing, apparently baleful look of a Martial Eagle's yellow eye as she stares at the hide in which I crouch, tensely waiting to press the shutter when she relaxes. No other bird has quite the presence of an eagle – not even a big falcon, such as the Peregrine or Lanner. They will keep me going as long as I can stand.

Eagle Hill, which has been for many years the central feature of my eagle studies, and is as a result known the world over, is in Embu District of Kenya. I still manage to be vague as to its precise whereabouts. It is not a magnificent mountain, but an upstanding one. On a clear morning after rain you can see all round for 160 kilometres (100 miles), save to the west where the towering snow-capped bulk of Mount Kenya hides all beyond. It juts out of heavily populated, intensively cultivated country and always has done for as long as I have known it. Where, in Europe or America, could you find such a mountain, with eagles living and breeding at peace, only a mile or so from thousands of the human race? Nowhere!

The total area of the hill is about ten square kilometres (four square miles), and it rises only about 460 metres (1,500 feet) from the surrounding rolling ridges. Composed of rocks of the basement complex, that is, the oldest underlying rocks in the earth's crust, it is not less than 600 million years old and probably older. It is the eroded remnant stump of a once much bigger mountain, standing far taller from the plain. The ancient rocks outcrop in several cliffs, the biggest about ninety metres (300 feet) high. We call this central cliff 'The Rock'; and I usually lunch there. Many big boulders, scattered or clumped in kopjes, stud the footslopes of the hill and its eastern end. The whole is now clothed in dense dark bush or savanna. On either side of the summit ridge are patches of cool forest, with trees about forty-five metres (150 feet) tall, and nice forest birds. At some past time two-thirds of the hill may have been forested, the rest being broad-leaved savanna or thickets of trees needing less moisture than true forest. There is no permanent water on the hill itself, but round its foot are springs that probably would not exist if it were not there, and two small streams trickle after good rains.

Eagle Hill was once inhabited by men, who cut down and burned most of the forest. The remnant patches are probably their old sacred groves. The men who lived there were moved off by one of my predecessors as Agricultural Officer in Embu District; I never found out who. It must have been about 1940–2 when the Colonial Government in Kenya finally woke up to the dangers of soil erosion on steep slopes, and did something about it. When I first knew the hill the bush clothing it may have been seven to ten years old, hardly, if at all, above my head. Now it is all some six metres (twenty feet) tall, and I must creep beneath it, for it is very dense indeed. Young trees, which will eventually emerge as dominants and make a true forest, push up through this dense bush, seeking light. One could accelerate their progress by clearing away the dense competing thicket, but I prefer to leave it to the natural forces of ecological succession. I have been watching it now for nearly thirty years and am just sorry that I did not realize at the time that I might still be watching it thirty years on and make more careful notes. No place on earth has such long-continued memories for me. Some of my father's ashes are scattered on top of the rock and mine will be too when I am gone. I have charged my successors with this duty.

I found Eagle Hill in 1949. It was part of a census area of 378 square kilometres (146 square miles) of Embu District, selected for its varied terrain, including cultivation, tracts of uninhabited savanna then rich in large wild animals, and a range of hills. Following upon earlier research of this sort in Scotland I was attempting a population survey of all nesting pairs of eagles. In Scotland, however, there is only one species of eagle, and with practice their cliff nesting sites can be pinpointed, provided no trees obscure the view. Here, there were twelve or thirteen species of eagles, and most of them bred in trees, scattered in uncountable millions over the whole area. I could not, in any reasonable time, search all this country alone; so I sought help.

It was ready to hand, in the shape of Wambere honey-hunters, each of whom had the right to hang his hives in certain tracts of the bush. They were completely illiterate bushmen, who would be despised by most educated Africans today. However, they were extremely knowledgeable about birds and beasts and could name most plants in their own language. Daily they scoured the bush looking at their hives, and they already knew of many eagles' nests and could find more. Keeping a jealous eye on their neighbours to see that they did not trespass, they would pass the time of day, and give news to each other. Accustomed from boyhood to climbing trees to reach their hives no ordinary tree held any terrors for them. Rhinoceros, buffalo, Lions, and other 'dangerous' big game that then abounded could be dodged or avoided if one was quick enough and kept one's eyes open. These men were made for the purpose, and willing enough to do what I asked.

I enlisted the help of these keen-eyed men by offering rewards, varying from five to ten shillings according to species and rarity, for a true report of an occupied eagle's nest. They were helped by small boys herding livestock to whom, in those days, five or ten shillings was riches indeed. So I soon began to get results; and after settling-down it went very well.

I had first to demonstrate to these honey-men that I would go wherever they could, was not to be put off by tales of ferocious rhinos or buffalo, and would pay only when I had checked the report and found it true. If I said I would be somewhere on Friday at four o'clock, there I was. Once it had been grasped that no wool could be pulled over this *Bwana*'s eyes (a reputation I had already acquired in my official duties as Agricultural Officer) many a small boy went away delighted, clutching five or even ten shillings in a grubby palm. The honey-hunters found it a very easy way of augmenting their income. In those days two good verified reports would pay their annual tax or buy them a good blanket.

During 1949 reports came in so thick and fast that I had to find a liaison man, to act as go-between and messenger between me and the honey-men to tell them as to where and when they should meet me to show me what they had found. At the time I used to hunt at week-ends with a man called Nyagga Mukinyu, whom you met in the chapter about Antbears. He was a good bushman and he produced for me another such, a little wizened middle-aged man called Njeru Kicho who became the best and most famous of all my eagle-watchers. Njeru was a honey-man himself and knew most of the others. For months on end most of my spare time was taken up checking on reports that he brought to me every few days. If it had not been for him, the others who were searching the bush might have lost interest, and the results would not have come so quickly. As it was, with Njeru as go-between and messenger, I was on the spot within a few days of any report, and could check its accuracy by sending him up the tree (if I could not climb it myself) to tell me what was in the nest. He very quickly grasped what was needed and for many years he was my right arm.

We first went up Eagle Hill in June 1949 and that evening found a Martial Eagle's nest, not used that year, and since then fallen down and deserted, though there is still a pair of Martial Eagles on the hill. Late in the evening, on our way down, Njeru showed me another great nest, not more than 800 metres (half a mile) in a direct line from the Martial Eagle's, which I therefore assumed was the alternate nest of that species. Njeru maintained that it was not, and that it belonged to a different species, even more 'maridadi sana' – very fine indeed – called a 'Koi'. I did not believe him; but of course he was right.

Two months later, I was sitting in the fork of a big fig-tree looking into this nest when the real owner swept past my face and alighted on a bare branch six metres away, full in the sun. It was a male Crowned Eagle, the most magnificent though not the biggest of African eagles, richly coloured, bold as brass, and reminding me suddenly of a pair I had watched for a brief and glorious day in Nigeria eight years before. The nest was huge; but Njeru said he could remember it since he was a boy, so it may then have been thirty years old. It does not look much different now, thirty years later, so it has been there for half a century or more, and must have been there when the hill was inhabited and cultivated.

We found nest after nest on Eagle Hill. In the whole of our census area we located, over four years, twenty-six pairs of ten different species of eagles. Of these, six pairs, of six different species, were on Eagle Hill. In other words, for some reason I could never fully fathom, almost a quarter of all the eagles were living on 3 per cent of the total area. At first I thought it must be due to the hilltop environment, a natural magnet for eagles hunting in the surrounding country. However, a careful survey of the other hills in the range included in our census area revealed only two other pairs in another 28 square kilometres (11 square miles) – equal to the average over-all density of a pair per 14·2 square kilometres (5·5 square miles). Again, three-quarters of the eagles nesting in the hills were living in about a quarter of the total hill area. It was, and is to this day, not easily explained.

The eagles on Eagle Hill included a pair of Martial Eagles, the largest in Africa; the pair of Crowned Eagles, the most powerful and only slightly smaller than the Martial; a pair of the magnificent, coal-black-and-white Verreaux's Eagles, about the size of Golden Eagles, feeding almost entirely on rock hyrax; a pair of African Hawk Eagles, for their size probably the most potent predatory birds in Africa; a pair of the then almost unknown Brown Snake Eagles; and finally a pair of the diminutive but spectacularly agile Ayres' Hawk Eagles, one of the rarest eagles in Africa, and at that time entirely unknown. Round the base of the hill there were several pairs of Wahlberg's Eagles and a pair of Black-breasted Snake Eagles, so that within 25 square kilometres (10 square miles) there were eight species of eagles. It was therefore natural to concentrate on Eagle Hill and its surroundings for the next few years, to make the detailed studies of breeding habits then entirely unknown.

Njeru's greatest find was the pair of Ayres' Hawk Eagles. We did not find them till 1950, and then nearly at the end of the breeding season, when the large young bird was about to leave the nest. The Wambere had names for individual species of eagles, unlike their neighbours the Wakamba, who call them all 'Ndiu' – just eagles – except the Bateleur, which is regarded as unique by every African tribe I know. Njeru came to me very excited and said he had found something entirely new – a bird he had never seen, did not know the name of, if it had one. 'It's like an *Njigi* (African Hawk Eagle) but smaller,' he said. I thought at first he must mean a Black Sparrowhawk, a forest species he might not have known about, but he scouted that. So I felt sure he was on to something good and went hot foot up the hill the next Sunday morning, when I had plenty of time. And, sure enough, there was the pair of Ayres' Eagles. Why we had not found them before was a mystery, for we had often been on the hill in the previous twelve months, and they are normally easy to see near their nest. However, it was the culminating event in a series of extraordinary finds, and Njeru got a double reward for it.

Of the six species breeding on Eagle Hill, only two, the Martial and Verreaux's Eagles, had ever been studied at the nest for any length of time. Verreaux's had been the subject of a classic study by E. G. Rowe in Tanganyika (Tanzania), at that time the finest and most detailed study done by anyone of any eagle. Rowe had mainly used illiterate Wambulu tribesmen as his observers with great success, and had worked out all the main details of the breeding behaviour of these magnificent birds. Apart from the fact that the pair of Verreaux's Eagles on Eagle Hill were breeding in a Euphorbia tree – the first known case of tree-breeding in a confirmed rock- or cliff-breeder – I did not expect that I should find out much that Rowe had not.

I myself had studied the Martial Eagle in 1949, at a nest on an escarpment near Embu where I built a terrifying 9-metre (30-foot) pylon, just balanced on the rocks by its own weight, and held in place by guy ropes. I took good photographs of that Martial Eagle from this dizzy structure and used Njeru to make more detailed observations, so that we recorded most of the necessary detail. I have not seen a better account in print to this day, though some additional details have been added. Unfortunately, my pylon was knocked down by Kikuyu to steal the planks that made its floor, and in 1950 that Martial Eagle's nest fell down. The tree was never used again, and no other nest we knew was so easy to reach and watch.

Still, round the base of Eagle Hill or on it we had five or six species whose nesting habits were entirely unknown in detail, though their nests and eggs had been described. I set out, with Njeru and various other helpers, to remedy this defect. In six months, by moving from nest to nest, we had amassed a great deal of new detail. Myles North, also an enthusiast of birds of prey, exclaimed only a year later, 'Goodness, you *have* been going it with your birds of prey!'

Between 1950 and 1952 I concentrated on learning all I could of the life histories of the various species of eagles in this tract of bush. Most of the detailed observation at nests was done by Njeru and various others whom he enlisted as assistants: Randani Gichimu, Ireri Mungato, and his brother Ngari Mungato. I found that the more illiterate and rough-hewn these men, the better their observations. If they wore smart clothes and did not smell strongly of a mixture of sweat and tobacco they were no good. Literate schoolboys were utterly useless, for they were not prepared to do the long hours of observation needed, nor would they push through thickets of thorn-bush which might harbour rhinos or snakes.

My illiterate honey-men all had sharp eyes and, being illiterate, they had astonishing memories. If you cannot read and write you must remember many vital details of your existence which a literate European would simply write down in a notebook and then forget. They could tell the time by the sun to within quarter of an hour or so, quite accurate enough for my purpose. As long as I kept in touch with them every few days, gave them regular instructions what to do, and wrote down myself what they reported, they needed very little supervision. Most of the detailed work, which was subsequently published in scientific papers and various books, owes far more to them than it does directly to me. I just organized it and checked up on what they said when I could.

Naturally, I tried to do as much observation as I could myself, not only because I wanted to check on my eagle-watchers, but also because I myself wanted to watch the eagles at close range. I was very soon convinced that, broadly speaking, what my men said would happen at a nest did happen. If they said an eagle ate lizards, or birds, or hyrax it did.

If they said that only the female incubated or fed the young, I found that correct. If they said that the male would bring prey between 11 a.m. and 1 p.m. the odds were strong that he would; and so on. I came to trust them and they me because, if I said I would be somewhere at a certain time, I was, and if there was any doubt about a nest record I paid a reward according to the evidence.

Sometimes they reported a nest with an egg or a chick which had disappeared before I could examine it. It would have been unfair on my part to assume that such cases were due to inaccurate reporting, for natural calamities do often occur. So, if it seemed to me that the signs at the nest bore out the report, I gave them the reward they thought was their due. Only a few cases of this sort occurred over three years anyway, and I usually found that the egg or chick reported was there in the nest.

Over these three years we gathered most of the unknown details of the breeding behaviour of the Wahlberg's Eagle, the Crowned and Ayres' Eagles, the African Hawk Eagle, the Bateleur, and the Brown Snake Eagle. We did comparative studies on the Verreaux's Eagle and some work on other species, including the Secretary Bird, which I regarded as a sort of terrestrial eagle. At many of the nests I had a hide built so that I could observe from close quarters and take photographs. I thought these necessary to illustrate books I hoped to write later to help pay for the considerable outlay on my part – for which I had no grant except a small amount from the Percy Sladen Memorial Fund.

Njeru and the others soon became expert at building these hides, which were often high in trees, sometimes 24 metres (80 feet) above ground. As far as possible we tried to build hides at nests outside our census area, so that the results there could not be affected by human interference. However, with one exception – the unique Bateleur – eagles are quite easy to watch and photograph at close range, so that we had little trouble of this sort. The main problem was that I could never make my Wambere grasp the fact that I weighed 90 kilogrammes (200 pounds) and was a good deal less agile than they were. Some of the ladders they built for me to reach hides high in big trees were terrifying. One, in a huge forest tree near Embu, from which I photographed another Crowned Eagle pair, led up a slanting limb to a hide 24 metres above ground. While climbing it I was attacked by the eagle, and though she did not actually strike she gave me the fright of my life. I hung on desperately, quivering with terror, until my heart steadied and I could climb the rest of the way. Then I took some good photographs, just to show myself I could.

Much of our detailed watching was done on Eagle Hill itself, and here we recorded the whole breeding cycle of the Crowned Eagle and the

Ayres' Eagle, while we watched Wahlberg's Eagle in the area at the base and the African Hawk Eagle on the next hill in the range. The Crowned Eagle was always my favourite and remains so to this day. They were magnificent great birds and, moreover, extremely bold, paying very little attention to us while we watched them from cover thirty metres away. I am not very imaginative when it comes to thinking up good names for my subjects, so I just called them Rex and Regina.

Regina was the most magnificent eagle I ever saw. She was much larger than Rex, who was no mean eagle in his own right; extremely bold, richly coloured, and with an impressive presence. When Rex brought prey she would seize it from him, mantle over it with wings spread, and yell in her loud melodious voice, glaring at him until he backed away and left. Then, in almost miraculous contrast, she would delicately and gently tear off strips of flesh and carefully feed her little eaglet. The contrast between strength and apparent savagery on the one hand and gentleness and care on the other was always entrancing.

In 1949, when we first found the nest, the Crowned Eagles laid, but failed to hatch. Njeru said that the nest had been raided by driver ants. In 1950 we observed the eagles from start to finish of the breeding season, recording the incubation and fledging periods accurately, which we later reinforced with more records. Crowned Eagles incubate for 49 days and the fledging period is 105 to 125 days, usually 105 to 115, averaging about 110 days. At the time these were the longest incubation and fledging periods recorded for any eagle, but since then we have found that much smaller Snake Eagles and the Bateleur have even longer incubation periods of over 50 days and very long fledging periods too. It was while we were concentrating on the Crowned Eagle that we found the Ayres' Eagle's nest in 1950. I was unable to do much serious work on them until 1952, as I was away in Britain on leave in 1951, and only returned in time to see the young Ayres' Eagle in the nest before it flew.

The Ayres' Eagles, quite different from the Crowned, were equally fascinating. They were small, only about a quarter of the Crowned Eagle's weight, incredibly swift and dashing, weaving in and out among tree branches at speed, combining the flying powers of a Goshawk and a falcon. If anything they were even bolder than the Crowned Eagles, attacking me without hesitation whenever I climbed into my hide overlooking the nest. It was most satisfactory to be able to record their breeding habits in full and photograph them at the nest in 1952, for apart from a few specimens and eggs in museums nothing at all was known about them. Crowned Eagles have been photographed by others, and studied a little elsewhere, but the Ayres' Eagles of Eagle Hill are still the only pair that has been studied at the nest in detail.

When I could spare enough time, usually at week-ends, my routine did not vary much. On the Saturday afternoon we would visit various nests that had been reported, or perhaps do a short spell of observation at one nest to record which of the sexes slept there at night – always the female in cases we observed, whatever the species. Then on Sunday I would leave camp early and climb the hill. In those days the thick bush swarmed with rhinoceros, and it was their paths that we used to reach the big rock. The bush was so dense, lacking any climbable trees, that had we been charged by an irate rhino there would have been nothing to do except wait till it was at point-blank range, and then dodge aside in the hope that it would blunder past and press on without stopping. Njeru assured me – from experience – that it would.

Fortunately, we had a redoubtable ally, my formidable white bull-terrier, Patchy. Like most bull-terriers he was brave and reckless to a fault, and many a hard run I have had trying – usually without success – to prevent him from hurting himself by attacking baboons or Wart-hogs, each of which wounded him severely several times. Of the three animals he could regularly catch – baboons, Wart-hogs, and rhinoceros – only the rhinos could not hurt him. He was completely their master, darting about barking madly in front of them as they made ineffectual pokes with lowered horns. He worked himself into an absolute lather of sweat until he finally exhausted himself, and left them to crash off on their way.

With Patchy along, we had no fear of the rhinos. We seldom ascended the hill in those days without encountering one or two, for there were at least twenty on the hill, as high a density of rhinos as ever known anywhere. The first intimation of their presence was usually a mad dash by Patchy into the thicket. There followed a chorus of barks and snorts, and then we would hear the rhino thundering away into the distance, pursued by Patchy. The barks and snorts would recede, and we would sometimes have to wait half an hour or so before we heard the panting breath of the dog as he unerringly returned to where he had left us. His sides heaving, tongue hanging out, he would look up at us and grin, as if to say, 'That *was* fun.' Even late in the evening we had no fear of rhinos as long as Patchy was around; he could cope with them, and did. So we could safely stay near the nest till dusk, watching an eagle go to roost, and then come down through the darkening bush, when we could hardly see, secure in the knowledge that no rhino would charge us suddenly.

By the end of 1952 I had gathered most of the more accessible facts about seven or eight species of eagles. The political storm was brewing in Kenya, and I had been transferred from Embu to Nyeri, to take over as Provincial Agricultural Officer. While this meant I could not continue with my detailed observations, I had done much of what I could do and could still go there at week-ends, while the huge area I now had to cover included the whole of the arid lands of north Kenya and the country of the pastoral Masai, so that my boundaries extended from Ethiopia to Tanganyika and included much wild and exciting terrain. I looked forward to continuing my studies in a modified way, but my superiors thought otherwise, and transferred me to Kisumu to take over Nyanza Province. I believe they thought they were doing me a favour.

In 1952 Regina disappeared from Eagle Hill. She had bred successfully in 1950 and had been on the hill in 1951, apparently hale and hearty, but in the breeding season of 1952 a new female appeared, Regina II. She was hardly bigger than Rex and was a duller, altogether less impressive bird than her magnificent predecessor. From that day on I have never seen another female as magnificent as Regina I. Something seemed permanently lost from Eagle Hill when she disappeared. I assume she died, as eagles must some day, but I shall never know how or where, though at least it would be reasonable to suppose that she had not been senselessly shot by an ignorant gamekeeper. My illiterate uneducated Wambere knew perfectly well that Crowned Eagles habitually ate a small antelope called the Suni, rock hyrax, and some monkeys, but were sure they never took domestic stock – just as they were sure that Martial Eagles did. And they were right.

When I had to move to Kisumu in 1953, more than 320 kilometres (200 miles) away, the Mau Mau emergency had already broken out. Restrictions were placed on the movement of Africans from Central Province, so that at first I could not take my trusted eagle-watchers with me. Later I managed to have them screened and accepted as uncontaminated by the Mau Mau, a movement which they rejected wholeheartedly. However, although we could watch other eagles, detailed work at Eagle Hill came to a definite end. I thought to myself that I had already achieved a great deal, and could not have everything; and set to work seriously on flamingos.

Despite the difficulties, I thought it would be worth while to try to maintain my continuous records on Eagle Hill, and I managed to do this with the help of my watchers. I could not always check whether a young bird seen by me had finally left the nest, but I accepted the reports of my men because I found what they said was accurate whenever I checked, and because they had no reason to lie about it anyway. They could not know when I might appear if they were in Mbere and I was in Kisumu. Later, when they came to Kisumu too, I was able to send them in advance to make checks and report to me.

Then in 1956 I was transferred again, as Deputy Director of Agriculture to Nairobi. Many people wondered why I would give up an interesting if exacting 'field' post at Kisumu for an almost purely administrative one. However, for me it meant that I could return at week-ends to my old haunts, and it also made my study of flamingos, then in full swing, very much easier. Flamingos were actually absorbing much of my spare time in those days. I visited Eagle Hill just often enough to keep up the records and try to make a film of the eagles. My Wambere returned home, stayed there, and reported to me at intervals.

The eagles had already been reduced from six pairs to four on Eagle Hill. The first to leave was the Brown Snake Eagle, which we studied in 1950, could not find in 1951, and which in 1952 and 1953 laid in new nests on the next hill in the range. This pair has never reappeared on Eagle Hill, though we often see them in the general area. Snake Eagles are different from most eagles in that they build a new nest almost annually, sometimes moving several kilometres from their old site. They are also irregular breeders, at least in Embu District. Their nests are hard to locate until they have young because they are small and slight, made of thin sticks, and in Mbere are almost always in the spreading crown of a Euphorbia, where they are very hard to see. However, if they had ever bred again on Eagle Hill we should have found them. They may easily have been breeding regularly elsewhere, but from now on I concentrated almost entirely on Eagle Hill.

The Verreaux's Eagles had also left by 1953. In 1951 their tree nest in the Euphorbia fell down, and in 1952 they balanced an insecure nest on a narrow sloping ledge on the side of a big boulder. What we thought was a new female laid one egg in this, but it did not hatch, and then the nest fell down. How they had got it to stick there at all is a mystery to me. They were still on the hill in 1953, but thereafter disappeared and, until very recently, never came back. Since 1975, however, we have again been seeing them on the hill, so they may try once more.

When I began visiting the hill more regularly again, in 1956 and 1957, there were still four pairs, and I decided that I would make a film of them if I could. I had already made some film of flamingos, and was intrigued with this new gadgetry. The Martial Eagle's nest was impossible for photography, being at the top of a tall isolated Croton tree with no other tree close to it in which a hide could be built. However, the Ayres' Eagle, the Crowned Eagle, and the African Hawk Eagle all could be photographed, though the Crowned Eagle was not ideal, as the nest was rather far away from the fig-tree on which we could build a hide. Njeru had failed to locate the new nest of the African Hawk Eagle; and I beat him to it. In a little forested gully, where we had found the first nest of the Martial Eagle in 1949, I saw a suspicious-looking dark lump in the middle of a leafy tree. I bet him the equivalent of a reward for this species – ten shillings at that time – that it was what he had been looking for, and unwisely he accepted the challenge. Sure enough, it was the African Hawk Eagle's nest, and occupied. I could not really make him pay, but made him work it off, which he did, over some months.

We built hides at all these three nests, but my film did not come to anything in the end. Firstly, I was very amateur, and took it all at too slow a speed; I could never have sold it to the BBC. Secondly, I gave the packets of film to an African messenger of the Agricultural Department to post, and it was only when a good deal of film had disappeared without trace that I realized something was wrong. Enquiries to Kodak elicited a denial that they had ever received any such film, and it transpired that the messenger had pocketed the cash and thrown away the film!

Thereafter I made it a rule to leave the office to post letters of any importance myself, relishing the fact that the Government thus had to pay for my expensive time instead of that of an African messenger. I salvaged some film, as a record, but I felt particularly savage about the loss of film of the African Hawk Eagle, taken in perfect conditions in strong sunlight. She had a small downy chick, which felt the heat, and struggled towards her. She did not move, but gently and solicitously unfolded one broad wing and held it as a parasol over the little chick, an utterly charming picture of a supposedly ferocious predator.

That pair of African Hawk Eagles deserted the hill after 1957. I could not explain why, as when the nest was last used she had reared an eaglet successfully. It was in a secluded place with no other eagle close by. However, she went, leaving three pairs on the hill, the Martial, the Ayres', and the Crowned. These three have since remained until 1976; and we have kept records of their nesting and success all that time. There is no other place in the world where this has been done for so long.

I was fully engaged on exacting official work, and on the study of flamingos, for some years. In 1958 I got married, built my house at Karen the following year, and my son was born in 1960. All these activities reduced my spare time, and also my spare funds, so that I could no longer employ my eagle-watchers permanently, and had to leave them in Mbere, using them from time to time. I still visited the hill each year often enough to check on what the eagles were doing; and the longer I kept on going there, the more I began to realize that such long-term records could be increasingly valuable.

In 1961 we had the Lancaster House Conference, and it became plain that we should all have to look for other jobs. I could have walked into a job with the World Bank, but that would have meant leaving Kenya and my house, which had cost me every penny I had saved over twenty years' service and could now be sold for one-fifth of what I had paid for it. So I decided that if I retired on independence, my golden bowler would just enable me to keep head above water while I made a living as an author and occasional consultant. Although, in 1963, no one could foresee what would happen in Kenya, we decided to stay on and stick it out as long as we could. We are still here and intend to remain as long as possible. At the end of 1978 I shall complete thirty years of continuous recording on Eagle Hill. The doings of the eagles there have been a continual thread running through the varied and, on the whole, quite exciting life I have led since leaving Government Service when Kenya became independent.

Visits to Eagle Hill in these later years followed a regular pattern. I used to go to Mbere and camp, meeting my eagle-watchers near their homes. Then I would climb the hill, spend a day on it checking up on what was reported or what was occurring, and return home the next day. As a free man I could now pick my times and dates, but although supposedly retired I was actually busier than I had ever been, travelling all over Africa for one reason or another, and going often to the United States and other places. All these other activities left me no time for detailed work on the eagles; but most of the detailed work had already been done in 1949–52, and my main object now became just to keep records of the eagles from year to year, for as long as I could. In 1953 I had already wondered whether it was worth persevering, but had done so with some difficulty and was glad I did. In 1963 I wondered whether I could carry on at all, and if so, for how long? However, all the time I kept up the records, and their value is now apparent.

The Crowned Eagle, the Martial Eagle, and the Ayres' Eagle continued to breed on the hill. The Martial Eagles were always rather shy birds so that we never could recognize them individually like the others. If they suspected that a human being was near the nest they just left; and we left too to make sure that their egg or chick came to no harm. The much bolder Crowned Eagles and the little fierce Ayres' gave us no such problems. We used to climb to the top of the main rock, from which we could look into the Crowned Eagle's nest below us and at the Ayres' Eagle's nest in the forest fifty metres behind us, and out over several thousand square kilometres of country in lower Embu and Kitui District. The Ayres' Eagles and the Crowned Eagles came and went in full view, undisturbed by us; and in the heat of midday we rested in the shade of forest trees. We thus had, and still have, a fairly regular and thoroughly enjoyable routine, which enabled us to record the doings of the eagles. As time went on it became, in some ways, more and more interesting.

When we began observing on the hill we had no ideas of the ages of any of the eagles nesting there. We just knew they were adults. Then the great Crowned female Regina disappeared in 1952. Her successor, Regina II, although a very much smaller bird, was mated to Rex. Rex himself disappeared in 1957, when we only saw Regina II at the site. Since Rex had first been seen in 1949, he must then have been at least eight years old as an adult. He was replaced by a truly magnificent male, Rex II, actually slightly larger than Regina II, the only case of this sort I have seen, for female eagles, like other birds of prey, are usually bigger and fiercer than their mates.

So now we had a male and a female Crowned Eagle, who had appeared for the first time as adults in 1952 and 1958 respectively, and whose entire breeding lives we had a chance of recording. Since there is general speculation about the age to which eagles live in the wild state (they can live to be forty or more in zoos) it was an intriguing prospect. The only problem was whether I should be able to keep records for long enough, and in the end I did.

We had an approximate record for the breeding life of one of the Ayres' Eagles in 1958. When I had resumed more regular recording in 1956 I saw that although the 1950 male was still present, the female was different. These birds were rather unimaginatively called William and Mary; and Mary II was a smaller, darker bird, with a paler yellow eye than her predecessor. Although I suspect that she had actually been present in 1955, I could not prove that she had arrived before 1956. When she disappeared in 1958, we knew that she had a known total breeding life of only three years, probably four. In that time she had laid an egg each year, and in 1957 laid two. The first of these failed to hatch, but she laid again in September, rearing that young one successfully.

In her last breeding season she was mated to a new and extremely distinctive small male, exceptionally pale in colour, and lightly spotted below. At first he was known as William II, as he replaced William I, known to have lived from 1950 to 1957, or eight years as an adult. In later years, as he went on and on, I called him Old Whitey. He was mated altogether to three different females. The first of these was Mary II, whom he inherited from William I. Then, between 1959 and 1963, he was mated to a particularly delightful little female, whom I named Quicksilver, because of the extreme agility of her flight in and out among trees. Quicksilver was very aggressive and would not hesitate to attack me face to face, which most eagles will not do – they prefer to attack quietly from the rear when they think you are not looking. Quicksilver lived till 1963, a total of five years as an adult, and then disappeared. She was replaced by a rather ordinary-looking female who, to our knowledge, never succeeded in rearing any young at all. Old Whitey continued mated to his third wife until 1968, after which we saw him no more. He thus lived certainly for eleven years and is the longest-lived Ayres' Eagle for which we have records.

The Crowned Eagles continued to breed every second year with a clockwork regularity which puzzled me. After the young bird had left the nest it was not practicable to search for it for long, and we had concluded that it soon became independent. However, in 1959, when I built my house at Karen, a pair of Crowned Eagles obligingly built their nest within sight of my study window. From daily watching of their comings and goings over the next few years we learned that the young Crowned Eagle is fed for between 9 and 11 months after it makes its first flight. With an incubation period of 49 days, and a fledging period of 110 days, this means that, if they are successful in rearing a young bird to independence, Crowned Eagles cannot breed every year, but only in alternate years. This seems to be true of Kenya, in equatorial latitudes, but South African ornithologists claim that in Natal the Crowned Eagle breeds annually, driving away the still importunate young one of the previous year when they want to start again. So far, the facts they have produced to support this are not detailed enough to convince me.

In 1961, when the Crowned Eagle's nest was occupied for a second successive year, I suspected at once that Regina II was gone or dead, and that a new female had replaced her as Rex II's consort. This proved to be correct. Regina III was a big, dark coloured female with very loose plumage, quite distinct from Regina II. Regina II had lived for nine years as an adult to our knowledge. In that time she had laid five clutches of sometimes one and sometimes two eggs, and had been very successful, rearing each of the subsequent young birds to fledging and probably all of them to independence except perhaps the last. This one would still have been dependent when she disappeared and Regina III appeared on the scene. Regina III would probably not have been willing to accept her predecessor's offspring and feed it, so it was probably forced to leave, though we never knew what happened. Anyway Rex II was quite content to take up with his new consort, and it became perfectly clear from these records that an eagle's nest is occupied, perhaps for many generations, by a succession of different birds. This explains why some sites are occupied for a century or more, though the individual life-spans of the eagles that breed there are far shorter.

The Ayres' Eagles had bred regularly, in the same nest, from 1950 to 1959, the last young one leaving the nest in January 1960, hatched from a replacement egg laid by Quicksilver. They appeared to be absolutely regular breeders up to that point. Then things seemed to go wrong for them. They did not breed either in 1960 or 1961, though the nest was still there, apparently undamaged. The pair were present, and they made sporadic repairs, but as far as we know did not lay. Njeru explained this by the presence on the rock of a pair of Peregrine Falcons, which he said so harassed the Ayres' Eagles that they could not breed in peace. This was perfectly plausible, for Peregrines do harass and attack other birds of prey or crows near their nests. They are so surpassingly agile that they could effectively harass even the Ayres' Eagles, magnificent fliers though these are. It may also have had something to do with, first, acute drought in 1960 and the first part of 1961, followed by tremendous rain and floods in late 1961. Or it may just have been Quicksilver's fault.

Quicksilver laid another egg in October 1962, in the same old nest. I had by this time acquired a 35-mm camera with a 300-mm lens and I was very anxious to take good colour photographs of the Ayres' Eagles. I was therefore very disappointed when this egg failed to hatch, and the nest was empty in December, when I could have spent a long week-end watching and photographing them. I told Njeru to keep a careful eye on them in case she might re-lay – as she had once before. It seemed very unlikely that she would, but it was just possible. It was Njeru's downfall.

I kept in touch with Mbere as well as I could, and Njeru reported that nothing was happening at the nest until July when, as expected, it was being built up. Since I was so anxious to get new colour photographs I went there as soon as I could and was assured by Njeru that there were green branches on the nest and that the birds should soon lay. This all agreed with past experience. However, I thought that, since I was there, I would check. Njeru seemed unaccountably reluctant, and my suspicions rose as I climbed the hill, finding our usual tracks overgrown, with no sign that anyone, let alone Njeru, had been there recently. He said, scowling, that they had gone by another route. However, he was just playing his luck, which was out. When we reached the nest, there was a great big young bird within a few days of its first flight. If it had been younger I could still have rebuilt the hide and photographed the parents, but it was now too late. Quicksilver had, after all, re-laid in March and had reared her young one successfully at a quite abnormal time of year, when all the eagles were normally inactive.

Njeru was thus branded unmistakably, as a liar; but he was simply furious with me for catching him out, and did not speak to me again for years. It was the end of an era for me, for thereafter I felt I could not trust anyone else, and that all future records would have to depend entirely on what I myself saw. As a matter of fact it was a typical piece of African behaviour: good and reliable, never faulted, for many years, they can suddenly grow tired of a job and give up. Usually they seem resentful that anyone should expect anything else, such as a word of warning that they are fed up with this job, and would you please get someone else to do it. They just stop doing it, no reason given.

Of course, Njeru's defection had nothing to do with it, but from then on nothing seemed to go right for the Ayres' Eagles. In 1964 Quicksilver disappeared, after a known life-span of five years, in which she laid four eggs in two years and reared two young, each from replacement eggs. The new female was a larger darker bird, with a very orange eye, quite distinct. She too was very aggressive, attacking me without hesitation. Half the nest had collapsed early in the year, but it was rebuilt, and she laid an egg in August. It did not hatch, and she did not re-lay, though I kept a lookout for this. In 1965 half the nest-tree fell down, probably because the weight of a creeper broke a big branch. This left the nest in full sunlight all day, and the eagles evidently did not like this and did not lay.

In 1966 the same pair was present, and they again repaired the old nest, but did not breed. In 1967 they finally abandoned this old nest, and it fell out of the fork by degrees. They laid an egg in a nest in the forest below the rock, belonging originally to a Harrier Hawk, and hatched their young successfully. There I photographed Old Whitey for the first and only time – not very successfully either. In November of that year this rather fragile nest collapsed with the young bird in it, so they failed to rear young. No young had now been produced by this formerly very regular pair since 1963.

Thereafter things seemed to go from bad to worse. They tried to build another nest on a lateral branch of the old nest-tree, but it fell down. They tried to build a new nest in a leafy tree below the cliff, but they evidently could not make the sticks lodge securely in the fork they had chosen. Thereafter, year after year, they built new nests in insecure sites and sometimes lined them, but failed to lay any eggs. They reared no young at all, and as far as I know laid no eggs, from 1967, until 1974. In that year a female, who is still present, laid an egg in a new nest, the third she had built, but failed to hatch it.

Old Whitey was not seen about the site after 1968. Although, in those years, I sometimes visited the nest site without seeing a male at all, when I did it was certainly not Old Whitey, whose snowy plumage was so distinctive that I would have known him anywhere. From 1964 to 1971 the same female was present, but she seemed an exceptionally incompetent bird, unable to build a secure nest, always choosing the wrong type of fork, or trying to balance her nest on a lateral limb. Her only egg was laid in 1967, in the Harrier Hawk's nest, and that failed. Although she lived for at least eight years as an adult, and is thus, apart from one other female at another site, the longest-living female Ayres' Eagle I have known, she left no offspring.

Meanwhile Rex II had been continuing, with Regina III, in the great nest below the rock. She was different from all other Crowned Eagles I have known, in that she was inclined to re-lay after a natural disaster. She reared her first young one in 1961, when she first appeared. She laid again in June 1963, as foreseen, but when that clutch failed, she laid another in November and reared an eaglet from that. Predictably, in 1964, she did not breed, but laid again in August 1965. This young one left the nest in January 1966, and was seen in September, just before it became independent. To my great surprise, Regina III laid again in late November, fifteen months after her last clutch, and the first time I had ever known a Crowned Eagle lay in consecutive years without the intervention of a natural disaster.

This egg resulted in a young bird still in the nest in January 1967; but it had died by February. Jeffery Boswall of the BBC and I found its corpse beneath the tree, which was distressing, as apart from losing the eaglet it was supposed to be a vital element in a film (*Birdwatcher Extraordinary*) about me, shown on the 'Look' programme in 1967–8. He thus had to manage without an eagle, and make do with some unusual starlings under a waterfall to vary a film mainly about flamingos and pelicans. Regina III laid again in September 1967, but late that year she went blind in one eye, cause unknown, and failed to rear that young one. I suspected that, with her handicap, she might have put her great claws through the eggshells. Anyway, she never bred again and disappeared in 1968–9. Rex II was there, alone, in 1969, but then he disappeared too. He was then known to be twelve years old as an adult and seemed perfectly healthy. He may just have left the site through loneliness, and not died, though that too seems unlikely, since he had been there so long.

In 1970 a completely new pair of Crowned Eagles appeared at the nest. The male was first seen in September, apparently alone. He was a very dapper-looking bird who, in addition to his official title of Rex III, was given the nickname of Smarty-pants, because of his neat, closely barred thigh and leg feathers. In October he was joined by a new female, Regina IV; she laid and hatched a chick late in December, which she then failed to rear. She did not lay as expected in 1971, but laid in August 1972, and this time was successful.

However, that was the last young Crowned Eagle to be reared on Eagle Hill. We saw the pair at the nest in every subsequent year up to 1975, but they never laid again. In 1976 I did not see them, but late that year they built up the nest with about 30 centimetres (a foot) of new sticks all round. Five visits in 1977 have revealed no new activity at all, so it looks as if the Crowned Eagles too are now fading out, for reasons we cannot explain. The great nest, which has been occupied since 1949 to my knowledge, and probably for sixty years, may soon be deserted and collapse. I hope not. Perhaps Smarty-pants will reappear, or another pair may take over, as he and Regina IV did.

During 1968 I did a resurvey of the whole of the census area I knew in 1949–52. Great changes had taken place in the land use of the area in those seventeen years. Almost the whole of the once uninhabited savanna, rich in wild animals, had been occupied and partly cultivated, and all but a few wild animals had gone. Even so, I found surprisingly little change in the population of eagles. We located a total of twenty-three pairs of eight species, as compared to twenty-six pairs of ten species in 1949–52. Possibly, given more time, some more definite pairs would have been found. The 'missing' species were Fish Eagles, both the nests known in 1949–52 having been since destroyed by cultivation along the river banks, and the Brown Snake Eagle, of which we could not locate a nest though we saw a pair. African Hawk Eagles and Bateleurs had unexpectedly increased, and Wahlberg's Eagles had surprisingly decreased from eleven pairs to eight. Best of all we found a second pair of Ayres' Eagles, a great find. In general, the re-count showed that a population of large eagles, in generally densely inhabited country, could still exist.

The same three pairs of Martial Eagles were breeding, one in the same tree, while the second pair had moved a kilometre or so from the 1949 site. The Eagle Hill pair were breeding a hundred metres from their old site, first occupied in 1949. We now have records for this pair of Martial Eagles extending over twenty-nine years. In that time they have laid nineteen times and reared thirteen young. The present pair do not seem to be able to breed successfully: they keep building new nests in insecure sites, which then fall down. They have not reared a young one since 1972, though the female laid an egg in 1975, in a nest built in 1974, but which she does not like – it is still there, unused.

Oddly enough, the only eagle on Eagle Hill breeding in 1977 was the Ayres' Eagle. The female at this nest, a very distinctive bird with large white spots at the base of her wing (hence called Whitespot), first appeared in 1972, when she built a new nest but did not lay. She laid her first known egg in her third nest in 1974, but failed to hatch it. In 1975 she did not breed, but built another new nest, and in 1976 laid again in this nest and reared the young bird successfully. In 1977 she laid again in this same nest, but has again failed. The nest, at least, has not fallen down. It seems specially curious that, after twenty-nine years, the only eagles breeding on Eagle Hill should be the Ayres' which for thirteen years previous to 1976 failed to rear any young and were only known to lay two eggs in all that time.

On 10 December 1968, during a night of very heavy rain, when, as one of my observers during that year's survey remarked, 'there were many evil spirits abroad', a landslide occurred on Eagle Hill. It started in the forests on the top, in four places. The descending slide of mud and rocks swept away big trees near the Crowned Eagle's nest and narrowly missed taking away that nest-tree too. It obliterated long-known landmarks such as a big rock polished by centuries of scratching by rhinos. The slide occurred between midnight and 2 a.m. and lasted about two hours, digging out a trench 27 metres (30 yards) wide and up to 4·5 metres (15 feet) deep, and, as one terrified onlooker observed, 'made a noise like fifteen aeroplanes'.

The path of this landslide is now our route to the top of Eagle Hill. I find it a good deal easier than creeping through thick bush, especially now that there are very few rhinos on the hill to open up the tracks. For a few years afterwards it seemed as though more people were going to the top of the hill, because of the relatively easy path that this landslide afforded. I had wondered whether the failure of the Crowned Eagle in 1970 might not be due to this. However, that was belied by the success of Smarty-pants and Regina IV in 1972; and the trench of the landslide is now overgrown again. I recorded the first plants that colonized the nearly naked soil of the landslide and amongst them found some tobacco, growing from seed left by those human inhabitants who had been moved off perhaps thirty years before.

I was going up this landslide in 1975 on one of my visits, when I became aware that two young men were following me. Ordinarily, I can evade people who want to follow me quite easily, just by dodging into the bush, which I now know far better than they do. However, these youngsters were too close, so when they came up to me I said that I did not want anyone with me. 'Yes,' they replied. 'We know who you are. You're Brown. But you're a *mzee* now (an elder – a term of respect); and we thought you might need some help.'

I assured them that I did not need any help, but that if they felt like coming with me they could do so on condition that they did not ask for any pay at the end of the day. They said they would come, and they had a hard day of it, cutting a path through dense thorn thicket at one point to reach the Martial Eagle's new nest. We reached the bottom again about three o'clock. By then I had learned that they were two of the growing multitude of hopeless school-leavers, with a smattering of education, but unable either to go further or get a job of any sort. One was the son of a man who had been an Agricultural Assistant in my time as Officer, and he cannot have been born when I started my work on Eagle Hill.

As we parted I thought, 'Now for the soft touch!' But they asked for nothing; just shook my hand, said, '*Kwaheri Bwana*' ('Go in happiness, Sir'), and went their way. Maybe they got something from a day spent in the company of this strange old *mzungu* (white man) who had been madly climbing that hill all these years, just to look at birds.

We already have records for twenty-nine consecutive years for the Martial and Crowned Eagles, and twenty-eight for the Ayres' – a rare species observed by no one else but me in any detail. We have recorded the complete life-spans of two female Crowned Eagles and one male, and of three female and two male Ayres' Eagles also, rather less accurately. Since I found Eagle Hill there have been four female and three male Crowned Eagles there and four male and five female Ayres' Eagles. The actual recorded life-spans agree quite well with predictions made many years ago, on the basis of the number of changes of mates observed. No such data have been gathered anywhere else in the world.

Eagle Hill is not a sanctuary, though it is a sort of forest reserve and so has some protection. I have never felt like pressing to have it made into a sanctuary because that might attract undesirable attention to it and because, as long as the Wambere who live round it leave the eagles alone, that suits me. So far they have always left them alone; in all my years of watching I have never once suspected that a pair of eagles failed because of deliberate human interference, though once or twice we have thought that people illegally cutting down trees to make beehives might have caused a desertion. That could not be said of any place or any tract of country that I know of in Britain, or anywhere else in Europe or North America. If there are such places, no records of them have been kept.

If I went in for such records I suppose I could say that I may have climbed Eagle Hill about 130 times. Allowing for 460 metres (1,500 feet) up and down (sometimes more, sometimes less) each time, that is about 60,960 metres (200,000 feet), nearly seven times up and down Mount Everest from sea level. I may have walked, or cut my way through thick bush, for around 1,290 kilometres (800 miles) in the process. I have myself changed from a tough and fit young man to an elder, who now cannot reach the top in much less than an hour and a quarter, whereas in my prime I could do it in forty-five minutes.

Well, I do not regret any of it. I am only sorry I have not been able to go there more. Sitting on the top of the rock on a fine clear day after rain, looking over a vast terrain, in which I have at one time or another climbed most of the other hills I can see, I look at country more full of memories for me than any other part of this world.

D. TISCHLER

6. Flamingos

Flamingos are the absolute antithesis of eagles. Eagles are wild, swift, generally elusive, always solitary creatures that often live in high mountainous country where one must be tough and young to make the best of them. They are, however, predictable. One may return to the same *Creag na'h Iolaire* (Eagle's Crag) in Scotland, thirty years after one last saw it, and there find an eagle breeding, as she always has done since time immemorial. Flamingos, in contrast, are intensely gregarious, living together literally in millions in East Africa. Apart from some remote Andean lakes, they usually occur in hot lowlands or sea-coast lagoons, where one can often watch them at ease without leaving one's car – though the Andean Flamingos at 4,200 metres (14,000 feet) are still one of the toughest ornithological challenges remaining.

Eagles may be spectacular and impressive, but they are not exquisitely beautiful and fragile, although slightly bizarre. Lewis Carroll chose right when he made Alice use a flamingo for a croquet mallet; an eagle would not do. And flamingos are utterly unpredictable and supremely opportunistic. Anyone who tries over a long period to rationalize what they do, is asking for trouble. One day they are there, in uncountable hordes, apparently restful and content. Next week they are all gone, flown out at night against the moon, who knows whither.

Even ringing, which has been very useful in Europe, has told us little about the movements of East African flamingos. Over 8000 were ringed on Lake Magadi in 1962, nearly all of them Lesser Flamingos. The most distant recovery was of a bird about a year old, which had flown to Sodere, in the Awash Valley of Ethiopia, about eighty kilometres (fifty miles) from Addis Ababa. All that it proved was that East African flamingos moved to Ethiopia at times, which seemed pretty certain anyhow. All other recoveries or sightings were from East African lakes, none more than a year old. We thought that the rings must have fallen off because of the chemical action of the water in which the flamingos lived. Then, in 1974, Chris Mead produced for me a ring from a flamingo eleven and a half years old, one of those ringed at Lake Magadi in 1962. The ring was in good order, suggesting that others could also have lasted that long. And where had that wretched bird chosen to die? Not in some remote lake which could have told us something useful about its wanderings, but at Lake Magadi, within a few hundred metres of where it was ringed! All that we learned from it was that flamingos could live for about twelve years, and that our convenient theory that the rings had rotted off was probably wrong. It could not have provided less useful information if it had tried.

In 1953, when I was transferred to Kisumu, kicking and screaming because I had been dragged against my will from my well-loved haunts in Embu District, I had to give up detailed eagle work for the time being. Though I returned to it later, and though I did make a study of Fish Eagles on Lake Victoria, I could no longer pursue my chosen occupation. I had a grinding load of work to do and little free time. However, over a long week-end I could now easily reach some of the Rift Valley lakes, where flamingos abounded. So I had the chance to solve what was then one of the most intriguing mysteries of the bird world – where did those millions of flamingos breed? No one knew! In 1951 Mackworth-Praed and Grant, the authors of the standard handbook of East African birds, could still write of the Lesser Flamingo, 'There is no bird of its size and numbers in accessible parts of the world about which so little is known.' They were right. So I set out to fill this gap, for it seemed to me that in a week-end I could certainly learn more that was new about a million gregarious birds than I could about solitary eagles, about which I already knew so much.

I had already learned a little about flamingos, for in 1949 I had visited Lake Hannington (now called Lake Bogoria). It was a light-hearted safari, and I did not expect to find out very much. I went with two friends, Tony Allison and Bill White, to the south end of the lake down the burning-hot, 600-metre (2,000-foot) stony and thorn-studded Ngendalel escarpment. It was a dramatic way to approach the lake, but a hard one, and perhaps a good introduction to flamingos and what their study might mean – a mixture of bitter hardship and transcendent beauty. On the way down that scorching cliff Bill White broke one of our two large flasks of water, and we feared the worst would befall. But we camped in a glorious grove of big fig-trees, at that time a truly enchanted spot, where there was a clear rushing stream. We cautiously made tea with it, though I have since learned that it is – or was – drinkable as it ran. It is a public picnic spot now, so I expect the stream is full of bilharzia. I have not been back.

I shot a duck for our supper that night, and had to swim in the lake to fetch it. The water was the vilest stuff I have ever entered, slimy with dissolved soda, pea-green and opaque with suspended microscopic algae, concentrated enough to make every one of the multitude of scratches and all the orifices of my body burn intolerably. When I put my feet down, I touched soft, sucking, impalpable mud, which I somehow felt was deadly and bottomless.

This, then, was the environment in which flamingos lived in millions, and from which they extracted their nourishment by a specialized filter in the beak that made it virtually impossible for them to live and feed anywhere else. I felt sorry for them; but in those days it was a grand, remote, and lonely spot. After a bath in the spring and a good supper of the duck's breast I rested content, on a luxuriant bed of star grass, out in the open, under the stars, where I always sleep better than under any sort of roof.

In the small hours I woke, conscious of a strange murmuring at the mouth of the spring. I sat up; and what I saw in the brilliant moonlight had me out of my bed and crawling slowly down to the lake shore. The flamingos were drinking at the source of fresh water; evidently even they could not manage exclusively with the foul waters of Lake Hannington. I reached a spot not more than ten metres from them and lay there for hours more, bitten by merciless hordes of mosquitoes, but totally entranced. In the moonlight, as they came and went, they had an ethereal pink radiance; and though panicking, sharp-sighted waders occasionally alarmed them when I slapped at a mosquito, they themselves paid no attention to me. They were exquisite; and were there because they needed a few sips of clean fresh water to drink. I had learned a little more about them before I crept back to my bed, and I can still see them in my mind's eye.

Nor was that all. In the morning the sun, that would grill us later as we re-climbed the scarp, rose over the cliff to the east. It lit first a long line of flamingos gathered along the shore of a bay on the far side, called Emsos. Then it picked out, as with a spotlight, one flamingo after another as they swam, feeding with a scything motion of the bill, over the glass-calm waters of the lake. Finally the whole bay was dotted with delicate beings like little pink swans, while masses still came to water at the outlets of the many geysers along the shore. I have many times since seen greater masses of flamingos, but nothing that I can remember made the same sudden impact of almost unbelievable beauty that those Lesser Flamingos made that first time I ever really looked at them. I knew just how foul was the stuff they swam in, but there they all were, fresh, delicate, crisply laundered almost, in their pink plumage. The 'little red water nymphs', *Phoeniconaias minor*: and still the greatest local mystery in the bird world.

When I began to study them seriously, it seemed that Lake Hannington was the obvious place to begin, for there, many years before, the

American hunter-naturalist and visionary taxidermist, Carl Akeley, had seen nests and found young. I did not then look up his original observation, but accepted what was repeated from book to book. What he actually said (as I later found) was ambiguous: 'the young birds in their grey plumage were abundant, and traces of nests were found at the north end of the lake'. It actually offered no definite proof of flamingos breeding there, but at that time was the only real clue, so to Lake Hannington I went.

At that time Baringo District, where Lake Hannington is situated, was what was called an 'outlying district' which one could only enter by permission of the District Commissioner, even if one was a Government Officer senior to him. I could never fathom the purpose of this regulation, which basically allowed certain District Commissioners to make themselves generally odious. All too often it gave them a bad dose of a disease I later called 'N.F.D. Swollen Head' (which did not add to my popularity). The man in charge of Baringo was one Simpson, and he enjoyed rubbing in this sort of detail. However, in 1953 I made a thorough reconnaissance of Lake Hannington with my father, who always enjoyed that type of safari. Some more of his ashes are scattered in the once enchanted fig-tree grove, and I hope he haunts the tourists with their Coca-Cola and transistor sets.

We camped first at a dreary place called Luboi, close to the mud-flats at the north end of the lake, where Akeley had found the nests. It was in March, and as Akeley had seen the young in May it seemed the best time of year to try. It was late in the dry season, a grim time of appalling drought, the thorn-bush naked, all grass long gone, and cattle dying everywhere. The veterinarian's needle had saved them from disease, only to die more cruelly later of starvation, after they had ruined their habitat. Luboi was a beastly place. Clouds of dust blew up every night, as the cold air from the high escarpments all round poured down into the heated trough of the Rift Valley. There were huge numbers of flamingos on Lake Hannington, and at the north end the flock varied from 70,000 to 120,000 on several counts. However, there were only a few score mud-mound nests and certainly no sign of large-scale breeding, so we had to look elsewhere.

We took porters from the village of Maji Moto and made for the lovely spring in the fig-trees called Mugun. They said it was only an hour and a half's walk, which seemed plausible; in fact it took us six hours. Since we had started late, the burning-hot ground blistered the feet of my poor dogs so that they were completely lamed. The temperature of stony ground in the sun in those parts can reach 60 to 65° C. (140 to 150° F.). We were told that the spring had dried up, but fortunately I did not believe my informants, and we then camped for four nights in the marvellous grove. We could sit in a cool bath of the dammed-up stream and watch the flamingos without moving. We called it ornithology *de luxe* – and it really was!

I spent my days walking round the lake, exploring any part of the shore that seemed a possible breeding ground. Although I did not realize it then, for I had been led to believe that Lake Hannington was their permanent stronghold, there was an exceptional concentration of something between 1½ and 2 million flamingos on the lake. Never since then have I seen such a multitude anywhere. The spectacle which had seemed staggeringly beautiful when we first came to Mugun in 1949 was now completely eclipsed, at least fivefold. We estimated that 80,000 watered nightly at the mouth of our spring alone, coming and going so quickly that each could take only a few sips before it was jostled out of place by the incoming horde. Even at the outlet of nearly boiling geysers flamingos drank almost all morning. The water was bitter, but not as concentrated as that of the lake itself, so it sufficed, even if it was very hot.

The greatest throng of flamingos, a flock about 720,000 strong, was gathered round the dry delta of a small seasonal stream. This place was called Gurasien, but I have also heard it called Legumugum, presumably because of the bubbling, boiling, sometimes spouting geysers that are numerous there. Later we gave these geysers names: 'the pipe', 'the bath', 'the death-trap' (for I do not think anyone could survive who fell into it), and others. Among the hot bubbling mud of some small geysers I found about 5,000 nests. They were empty and could not possibly have produced the 150,000 young flamingos in grey plumage present on the lake, but at least they were a clue.

When we left Mugun, having done all we could reasonably do without any definite result, the dogs' feet were still bad. Half our porters also failed us. So I walked back at night through the moonlit bush to Maji Moto, while my father at dawn led the safari up the cliff. Not knowing how to carry the dogs, he had the brilliant idea of causing a porter to sling each dog round his neck, as a shepherd carries an orphan lamb. I met them all safe and sound, some hours later, at the top. Years later, this exploit was still remembered by local people. The old mad *Bwana* who cursed and goaded the porters to carry dogs up a steep hill was unforgettable. Perhaps his ghost does it now; it is the sort of thing his ghost would do.

Since I still had no better clue, I regularly visited Lake Hannington for a year. I now based myself on the Maji Moto (Hot Water) spring, which became a haven of rest and peace for me. It was glorious to arrive there late on a Friday evening, certain of two good days free of paper and telephone calls, and lie for an hour in the warm bubbling water of the spring before drinking a bottle of beer at peace beside my camp fire. Sometimes my father came with me, or other friends. There were frogs in the spring which uttered a reedy chorus, interpreted by my father as 'Fresh! Fresh! Very-fresh!'; and it was. In those days we often heard the grunt of a Leopard as it came down the valley. The bush abounded in dik-dik and guinea-fowl, and there were nice things to see when I could take a leisurely evening stroll. Before I left I usually used to shoot a brace of guinea-fowl early in the morning before breakfast. One day I had an unexpected bonus from a male Chanting Goshawk, who killed one just across the spring from me. He ate only the flesh of the neck, and I took the rest.

To reach Lake Hannington from here I had to take my Ford pick-up a kilometre or two over two stony ridges, christened by my father Bugger and False Bugger, on the ground that anyone sitting in the back of the car would be unable to refrain from that unnatural feat! I left it in a shady acacia grove and then walked 8 kilometres (5 miles) down over more stony, thorny ridges to the Gurasien Delta. I had concluded that if flamingos bred anywhere on Lake Hannington it would be here, though as a precaution I usually checked the Luboi flats too. I would spend the day there, resting at midday under big Acacia trees, where Tawny Eagles had perched and plucked the bodies of flamingos they had killed or scavenged. Then, when the first scorching bite of the sun had gone, at about three o'clock, I used to start back. It was always a gruelling trek, and sometimes, if I had been energetic along the lake shore, it became a killing struggle. Those who now drive down there in an ordinary car to view the flamingos and the geysers in what is to become or is already a National Park, cannot realize what it was like then. The few score kilometres and the few hundred metres I walked up and down that path seemed much worse to me than all my climbing on Eagle Hill.

Flamingos did indeed make nests at Gurasien, but never in the enormous numbers that could explain a population of several millions. Nor did these nests ever come to fruition. I found a few eggs, sometimes even laid on top of a typical mud nest-mound, but they were never incubated or hatched. In October 1953 I found several hundred nest-mounds, and the flamingos were behaving as if they were about to lay. However, by November they had abandoned that colony and started another. That too was abandoned, but at Christmas a really solid-looking colony of about 3,000 nests had been built out on the mud, about a hundred metres from the shore among the bubbling hot outflow of the geysers. In earlier colonies of nests many of the birds had not been in full pink breeding plumage, but all these birds really looked like full adults. I thought I had it at last.

Next day I came back with a Tugen elder called Kipkwe, who had been with me several times before. He was a good and knowledgeable bush companion, and he told me that when he was young, maybe twenty-five years before, the barren scorching bush we now traversed had been rich grassland dotted with tall acacias, with rhinos and large herds of Eland. Through overstocking and misuse it had since deteriorated to bare stony ground with thickets of uneatable bush and succulents; but he could not connect cause and effect, and blamed it on God ('Shauri ya Mungu, tu!').

He installed me in my hide and retreated to rest in the shade while I sweated in my hot little canvas box. I had forgotten to take any drinking-water into the hide with me, and when I came out later gulped a deep draught from my one-gallon plastic bottle. It produced an intense desire to urinate – which I tried to, but without success. Kipkwe laughed. 'Kisonono cha kiu!' he cried; which literally interpreted means 'thirst gonorrhoea'. I had dehydrated myself to the point at which I had no surplus fluid to void, and very uncomfortable it was till the water worked its way through my system. However, I had taken the first good photographs known of Lesser Flamingos on their nests, and that was something, even though they had not laid any eggs.

I asked Kipkwe whether he had ever seen young flamingos produced in that place. He regarded me with an air of avuncular disdain, and said: 'You say that these birds lay eggs on those little mounds, and hatch them like chickens. However, *we* know that they don't do it that way. Their young are produced, nearly full grown, and able to fly, in hordes, all of a sudden. They don't breed like other birds at all.' I thought he was talking rubbish, of course, and told him so. Little did I know that he was, from his viewpoint, giving a plausible explanation of a phenomenon he could not explain, or that I, in trying to explain it, would more nearly meet Lucifer face to face than at any other time in my life.

I could not follow up the colony of 3,000 nests, for shortly afterwards I had to go to Britain for medical treatment. It was only a small skin cancer caused by exposure to the sun; I have had many more since then. However, I was told I must go, and spent six weeks in Britain, during which time I felt sure those wretched birds would breed, and beat me to it. Fortunately, another flamingo enthusiast, Lord William Percy, was able to help, and he twice made the arduous journey through the thorn-bush to Gurasien. He found nothing – the nests all abandoned, no broken eggshells, nothing to suggest breeding at all. Even most of the flamingos had gone, a fact I confirmed for myself as soon as I could. The great flocks at Gurasien, regular for more than a year, had just disappeared into thin air. Evidently they must have gone somewhere – but where? Lake Hannington certainly was not their main breeding place, anyway.

In early 1954 the drought broke, as droughts do, in torrents of rain. Lake Nakuru and Lake Elmenteita, which had been nearly dry, filled up again. Since most of the livestock had died in the drought the bush near Maji Moto bloomed and blossomed, the grass sprang up miraculously, and what had been an arid thorny desert seethed with life. It was my first real lesson in the astonishing powers of recovery of the African bush, after what appears almost irrevocable misuse. To be sure, much rich soil had gone for good, carried into Lake Baringo in the first violent floods, but now everything was growing again. I had a delightful safari in June with a friend, Charlie Hill, but we found few flamingos, and no sign of large numbers at Gurasien.

The few that were there were performing their abandoned, crazy-looking nuptial display. Although this has now been given a better-sounding scientific name (the 'tall strut'), I called it then, and still call it, the 'communal stomp'. Packed together, body jostling body, a group of about ten forms up behind a leader, whom I now know is usually a receptive female. She erects two bright-red tufts of feathers on either side of her tail, to indicate that others should join her in the dance. First ten, then a hundred, then a thousand, and finally – if there are enough – many thousands join in, till the packed mass marches insanely through more scattered flocks like some composite, multi-legged monster. It emits a sound like the roaring of distant surf on a bar, compounded of squeals, grunting murmurs, and the splashing of many feet in water. Within the mass some birds point their beaks skywards, waggling them frantically from side to side. Others place their beaks against their breasts, in an unnatural stiff attitude I call the 'broken-neck posture': it seems as if the bird has broken its neck a few vertebrae behind the head. Yet others bicker constantly with rapidly fencing beaks. The feathers of the head and neck are erected, and I fancy there is a flush of red blood to the skin. The rushing, surging mass of pink bodies is thus surmounted by serried ranks of dark-red columns, the whole carried along, as if involuntarily, on a forest of twinkling red legs. Absurd indeed, but eye-catching! To me they seemed to mock at my efforts to understand them, and to say: 'Oh, yes, we're going to breed somewhere, all right!' Well, I was soon to know where.

Having drawn blanks at Lakes Hannington, Nakuru, and Elmenteita, I had eliminated the easily accessible places. There remained, far to the north, Lake Rudolf (the Jade Sea), the largest alkaline lake in the world, stormy and unpredictable, not to be tackled lightly. To the south were lakes Magadi and Natron, almost solid expanses of crystalline soda. Lakes Balangida Eidahan and Lelu, which Reg Moreau thought might be likely places, and Lake Eyasi, remote, wild, lying by itself in a separate trough of the Rift Valley. Even to reach its shores would be quite a safari. If all these failed, then there was Lake Rukwa, far far to the south, in those days almost weeks away by car from Kisumu. They all had to be eliminated, bit by bit, now that the easy nearby ground had failed.

I took stock of the situation carefully, looked at what maps there were, and calculated that to walk round Lake Eyasi alone, supposing that I could obtain the necessary porters for the safari, would take at least ten days. I had an exacting job to do, and was none too fit. I would have been driven almost insane by frustration if I had found a colony late in such a safari, having used up all the leave I was entitled to, and then been unable to watch it come to fruition. Then it struck me that there was one obvious, if rather expensive, way out of the difficulty – an aeroplane! I wondered, later, why I had not thought of it before, but in those days we were not so accustomed to small handy aircraft for wildlife surveys as we are now.

I tried first to get an amateur to fly me, as the cost was really quite daunting for an impecunious Government Officer. In the end I had to hire an aircraft, piloted by Z. Boskovic, 'Bosky', a famous Kenyan character now, who runs his own airline. Since there was a spare seat I took along Peter Bally, botanist at the East African Herbarium, and a reliable scientist who could confirm what, if anything, we found. We planned a route to include all the likely lakes in north Tanganyika and southern Kenya, leaving the circumnavigation of Lake Rudolf in a boat I did not own, and a lengthy safari to Lake Rukwa, to the category of desperate last alternatives.

We took off at 15.30 hours on 20 August 1954, less than eighteen months after I had started on my study of flamingos. Twenty minutes later we had flown over Lake Magadi, and eliminated that – always rather unlikely. We went on over Lake Natron, but found nothing in what seemed the only likely spots, at the mouth of the southern Uaso Nyiro and the Peninj River delta. Then we flew on making for Mount Gelai, round the edge of a tongue of evil-looking pinkish soda mud-flats protruding into the south lagoon. I suddenly saw a dark patch in the water; and conviction was instantly born. They were a pack of the hitherto unrecorded, unseen, unknown young of the Lesser Flamingo, right out in the middle of this scorching, stinking mud-flat, almost the last place where I would have looked for them.

There were only about a thousand, and we had little time to search if we were to complete that day's flight plan. However, we found many nests scattered over the cracked, crystalline surface of the mud. I at once concluded, wrongly, that this was why no one had ever found them before – they must breed scattered, in ones and twos, all over such vast expanses, not in big groups at all. We saw no more young, so we turned and went on, over the wonderful, jewel-like crater lake of Empakaai, and past the great mountains of the Crater Highlands, Loolmalasin, Olosirwa, and Ol Donyo Lengai – God's Mountain, still an active volcano. It was, and still is, one of the wildest and most magnificent parts of Africa. We overflew Lake Manyara, now a National Park made famous by Douglas-Hamilton and his elephants, without seeing any nests, and landed at Arusha for the night.

Next day we flew over the likely Lakes Balangida Eidahan and Lelu and the remote Lake Eyasi, with no result. Lake Eyasi, which I had thought most likely of all, was a largely dry expanse of dust, blowing into dunes. So evidently it was necessary to search Lake Natron more thoroughly on the way back. We flew straight to where we had seen the herd of a thousand young, at the edge of the mud-flat. There they were – about 3,000 now – and behind them, stretching into the distance, a long black line of thousands more, gathered into little knots here and there, but all trekking, alone and unaccompanied by adults, across those frightful burning-hot soda-floes. Following the line we came to huge empty colonies of nests. The great string of young we had seen must have left the colony only that day, and had trekked, unguided, for many kilometres in the blinding heat of the midday sun, with no fresh water, and going they knew not whither, to join the other birds at the toe of the mud-flat.

A few minutes later we found another colony, of at least 50,000 nests, still with eggs, and young too small to walk. Among the great numbers of Lesser Flamingos were several groups of Greater Flamingos, proof positive that they too bred in this fearful place. Peter Bally was busy with clicking camera; and Bosky flew us brilliantly, quite unconcerned, to and fro and round and round till I felt I had recorded in my notebook all we were likely to see. Without a doubt we had found one, if not the main or only, breeding ground of the Lesser Flamingo, and we could have done it within an hour of leaving Nairobi if we had known. It was, still is, the most dramatic and exciting half-hour I have spent. We landed soon after at Nairobi and my cup was full. It had been well worth the seventy pounds it cost – chicken-feed to any of those well-funded Ph.D. scientists today.

* * *

Well, now I knew. The mystery was solved, the flamingos who had kept their secret so long were beaten by modern technology. But, how to get there – that was the problem! I had estimated that between 100,000 and 150,000 pairs of Lesser Flamingos and about 500 pairs of Greater Flamingos were nesting out on that soda-flat and that the colonies were at least six and a half kilometres (four miles) offshore. I now believe that I underestimated the numbers, and I know I underestimated the distance, for the colonies must have been some thirteen kilometres (eight miles) from shore. However, this was my chance; and seize it somehow I must.

It was ten days or so before I could escape my office, with a few days long-due local leave. I drove to Magadi, and from there along a very rough and barely perceptible track to a magnesite outcrop at the foot of Mount Gelai. Here I camped; and next day set out, with my trusted Mbere eagle-watcher Njeru, to reach the colony. I had taken what precautions I could. I had plenty of water, sun-goggles, and a pair of mud-boards for my feet, in case I sank in. I had asked the medical men whether the soda or the noxious gases that rose from the mud-flat would harm me, and had been told that, if anything, the soda was good for sores! So, having found what I thought was the right acacia tree, I set out quite confidently, on a bearing that would take me to the invisible colony, across an arm of shallow stinking water in which the bodies of many dead locusts floated.

I found the mud-boards useless at once, discarded them, and continued in gumboots. I was a practised walker in mud, having spent many winter nights walking the mud-flats of northern estuaries. I still believe that this single fact saved my life. It was heavy going across the shallow water, but I expected to emerge on a dry mud-flat with a hard crystalline cap, on which I should be able to walk fast, and reach the colony in two hours or less. How wrong I was!

I emerged on to the drier mud-flats beyond to find that, instead of being easier to walk on, they were much more difficult. The crust of crystalline soda would not bear my weight, and I cracked through it at every step, to sink slowly into glutinous, black, stinking mud. I was forced to incessant racking effort, for I dare not let one boot sink in so far that I could not then withdraw it. In northern estuaries, in such circumstances, one lies down and rolls out; but I did not fancy rolling for several kilometres in that heat. So I struggled on, growing more and more exhausted with each step. I have since taken the temperature of those flats, and know that it sometimes reaches 74° C. (165° F.), much hotter than you could bear your bath. The high temperature and the blazing sun were dehydrating me very rapidly. I took a swig from my canvas water-bag – chosen because it was unbreakable. Horrified, I found that it was already impregnated with soda, so bitter as to be almost undrinkable. So I drank most of it before it could get worse. A young flamingo ran mockingly, and without difficulty, over the soda crust in front of me.

At length, after struggling on for perhaps a kilometre, I climbed out on a small soda island, cracked through with both feet, and came to my knees, utterly done. I realized not only that I had not the slightest hope of reaching the colony, still not in sight, but that I would have to husband every bit of remaining strength to escape alive. I was much worse off than I knew, for not only had the extreme heat dehydrated me to a dangerous degree already, but chunks of the solid soda had fallen into my gumboots and were lacerating my feet and ankles at any point that could rub. I paid little attention to this at the time, reassured by what the medicos had told me.

Buoyed by hope, enthusiasm, or satisfaction after achievement, it is extraordinary what one can do. But the mountaineer who fails to reach the summit of a high peak finds the descent harder than he thought. So it was with me. Crushed now by disappointment, I had to make my way back across a kilometre or so of evil, stinking, soda-flat, and then across the arm of shallow water to the shore. I was utterly exhausted and dehydrated worse than I knew, but still must somehow put one foot in front of the other and take a few tottering steps before being brought up short, gasping for breath, as one boot stuck firmly. I was then forced to embed the other more deeply to drag it out. I dare not stand and rest, for that would mean both boots firmly stuck, and me rooted to the spot until I died. I could see it was a beastly way to die, like those locusts embedded in the filth; and it was partly this thought, and partly sheer anger at my own stupidity, that kept me going.

So I began the dreadful treadmill. A few tottering steps; stick; gasp; wrench out the foot somehow; and totter on again till the mud once more grasped me. I was so far gone that the huge, 3,000-metre (10,000-foot) bulk of Mount Gelai, only a few kilometres away, faded from view, and I lost the clear black trail of my own footsteps which could have guided me. Somehow, I did not go round in a circle, which would certainly have been the end of me. By sheer luck I suddenly struck a patch where the going was a little easier. I took twenty steps, struggling and gasping. But this break in my enforced rhythm of struggle had given me just a little needed respite. Although I do not generally believe in such things it seemed to me at that moment that some Providence lifted me bodily and helped me through. It came to me that, though I could almost see the Devil eagerly waving his pitchfork, my time had not yet come. I struggled on; and I made it.

Shortly, I came to the arm of water, where the going was wetter, so that it was easier to withdraw the stuck boot. I could often take twenty steps together and then allow myself a little more time to gasp between bouts of effort. That certain twist and pull, acquired in winter nights far away, in cold wet glutinous mud, stood me in good stead now. I got out, and rested finally on the hard soda-flat near the shore.

I withdrew my lacerated feet from my gumboots. They were an awful sight, covered with great dark-red blood blisters which, as I watched them, turned slowly brown and then black as they dried. I sloshed the little bitter water I still had over one foot. Then I somehow crammed them into a pair of light shoes and walked back to the acacia tree where I had left Njeru, washing my feet and legs again in a relatively fresh spring on the way. Njeru had already drunk nearly all of what I needed most – the spare water. He could have done nothing to help anyway in his bare feet. He said: 'I saw you standing still a long time; and then you turned back, and I knew it must be bad, for you do not easily turn back.' I lay down under the tree, dead beat and in a much worse state than I realized then.

When the sun cooled a little we returned to camp. I drank and laved my body in a not too bitter spring. I felt better, invigorated and rejuvenated after restoring water to my dehydrated body. At camp I disinfected and powdered my feet and actually felt equal to making another attempt on the morrow. Two magnificent male Lions were roaring at each other across the little valley, and I hoped they would go away, for a Lion roaring in your ear does not encourage sleep. They did, and I slept, exhausted; but not for long.

I was woken about two hours later by intense burning pain in my feet. I stuck them out into the moonlight and saw they were hugely swollen, oozing blood and pus round the edges of the blisters. They had gone septic. The problem was to reach the only place – Magadi – where I could obtain treatment. I had to make the agonizing decision whether to start now, perhaps lose the indistinct track in the dark, and thus be finished off completely or suffer the pain somehow and wait till dawn when I could see. Reason dictated accepting the latter unpleasant alternative. By taking three Soneryl and four aspirin tablets, I suffered through the night in fitful drowsiness.

At dawn I had to drive the 64 kilometres (40 miles) to Magadi. It was not too bad at first, for my feet seemed to hurt less and I was still drowsy with the drugs. Then the sepsis grew worse again, and the muscles of my face contracted, so that I thought I had lockjaw. Worst of all, my eyes would not stay open, and I finished the journey into Magadi holding my right eye open with my fingers and changing gear with my left hand. On such bush tracks changing gear is repeatedly needed. Each time I had to summon what courage remained to make the agonizing effort of depressing the clutch. I was now absolutely certain that I would reach help, only to have my feet amputated. I could not see how they could be saved.

Suffice to say that, after a five-hour journey, I reached Magadi at lunchtime. The young European policeman from whom I begged help would not rise from his easy chair to give it, and the Asian doctor to whom I eventually went at first sent a note to say that he would see me when he had finished his lunch. In agonized fury I struggled out of the car and thrust my bursting bleeding feet between him and his curry. When he saw how serious it was he was kindness itself, and he undoubtedly saved my feet. Morphia and soothing lotions soon brought forgetfulness.

I lay in Magadi for a week, gradually recovering, and then returned to Kisumu. I rang up the medico who had advised me that the soda could do no harm, said I had something to show him, and asked him for a drink. When he came I showed him my feet, still covered with open though no longer septic sores. Eventually I spent six weeks in hospital having extensive skin grafts which healed the sores; now I can hardly see where they were. But I have not forgotten what it was like, and when I see some young fool walking about on those soda-flats in tennis shoes I tell him 'Don't'. They usually just smile and ignore me; they know it all, of course.

After this almost mortal blow it was a long time before I could do much more about flamingos. I felt that since I had got myself into trouble by my own stupidity, I was duty bound to give up all my local leave and get on with my work. Had I known what Mr. Macmillan and his Colonial Secretary, Macleod, had up their sleeves for Kenya I doubt if I should have been so conscientious; but I did not know, nor did anybody else. Then, in 1956, I was transferred to Nairobi, as Deputy Director. This meant not only the opportunity to return at week-ends to my old eagle haunts but much greater opportunities for studying flamingos. Work to me is something I do for pay; if I like it, that is a bonus. It is of much lesser importance to me than how I spend my free time.

The realities of Lake Natron made it clear to me that I should not achieve much there without special equipment – which I could not afford. However, in Nairobi I could learn to fly an aircraft, which would mean that I could then survey the likely breeding grounds in two hours of an evening, after work. The true flying men, who were interested in flying for its own sake, thought it odd that anyone would learn to fly just to study flamingos. However, it was also very enjoyable, and I found it a useful accomplishment in many other ways. One can see so much from a small aircraft which is not easily appreciated on the ground among vegetation: the shapes and forms of erosion on earth, and what is happening in a place it may take days of foot-slogging to reach, and which therefore no one visits. Nowadays most wildlife biologists rightly feel that they must fly; I was one of the earlier ones; but no one paid me a penny for it.

In the meantime I filled in details. I read widely about flamingos when on leave in 1955, and even contemplated trying to do a Ph.D. on them, but concluded I would not have the time. Within three years of beginning the study I had gone a long way. I knew now not only where the Lesser Flamingos bred, but also that Greater Flamingos were not just visitors, but bred here too. Soon these magnificent great birds were to replace the Lesser Flamingo as the main stream of my study. Having been within an ace of the Devil's clutches, I was now to have a year of almost pure ornithological feasting, an experience I can never repeat.

In October 1956 I went to Lake Hannington with my father on one of our periodic checks. We had to pass Lake Elmenteita, and I noticed that there were pale pink masses on some of the islands near the far shore. I guessed that they were Greater Flamingos, but hardly dared to think that they could be breeding there, for the islands were hard lava rock, on which no flamingo could make a typical mud-mound nest. However, my diligent reading in 1955 had included an account of Chilean Flamingos nesting on a rocky island in a South American lake. I had done my homework; and as a result, when there was nothing much happening at Lake Hannington, we checked those islands on our way back. And there before us was the greatest spectacle in the bird world – a colony of Greater Flamingos in full breeding cry. The unbelievable was true.

They were on islands a kilometre out in the lake, and the first need was a boat. Fortunately, I had just what was required – a light two-section plywood dinghy bought for trout fishing. We launched it on the lake and rowed out gingerly, expecting the birds to panic at any minute. However, they did not, and we were able to feast on the spectacle at a range of not more than twenty metres without greatly disturbing them. Several of the colonies had already hatched, and there were little chicks everywhere. The babel of trumpeting voices, the great long legs and serpentine necks, the pink plumes raised in territorial display, the brilliant colour and animation of the scene were all almost more than one could comprehend. And it was supremely beautiful, like nothing I ever saw before. I have seen it again many times since, but on that first occasion the impact was stupendous.

I concluded that I could not fix any ordinary hide to those hard black rocks, and wondered how I could approach really close. I could creep up to a distance of ten metres or so in a boat, but once the chicks were mobile they were soon led away by their parents, amid a chorus of alarm. It occurred to me that if I made a hide floating on a pontoon they might not object. So I designed a structure based on a rectangular sheet of plywood, supported on four galvanized drums, which I still have and use. It took a little time to make, and in the meantime I kept on watching the colonies, taking other people to look at them and take film.

At Christmas I had a long week-end free, and there were still several islands with eggs when I arrived at Elmenteita. I thought I had plenty of time in hand to give the hide a thorough try, and perhaps achieve my aim of making a film and getting really close to flamingos. I had timed it perfectly, for the chicks were about to hatch. However, I had reckoned without Nature, whose quirks are often very strange.

On the evening I arrived, so did six Marabou Storks. Why they went straight to the islands on which the flamingos were still sitting, I cannot to this day understand, though I have seen the same thing happen again several times since. However, choose those islands they did, and in twenty-four hours or less they caused all 700 remaining pairs of flamingos with eggs to desert. Though there were only six of the hideous great storks the flamingos just did not seem to be able to stand up to them. In the morning, with all my hopes dashed, I found that many deserted eggs had hatched into adorable little gosling-like creatures, with dark eyes, soft silky otter-like down, and swollen pink legs. I took several of them for the museum, but I was choking with fury at the Marabous. Just another twenty-four hours and all those beautiful little chicks would have hatched safely, and many would have lived.

It seemed that I was to be deprived yet again when success was within my grasp. However, once more I reckoned without the capriciousness of Nature. No species of flamingo had ever been known to breed twice within a year; indeed they are usually very erratic breeders, not breeding for several years and then suddenly breeding in enormous numbers together. I expected nothing, but since I had to travel regularly along the road to Nakuru, I kept an eye on Elmenteita. Greater Flamingos persistently performed their nuptial displays on submerged shoals during March 1957 and they sometimes came out on bare rocks and put their heads down, as if to nest, fluffing their plumes in what I knew to be territorial display. I scarcely dared to hope, but the very unpredictability of the flamingo prevented me from wholly giving up. They *might* breed again – what was there after all to stop them if they felt like it?

On 14 April I had to attend a Pyrethrum board meeting at Nakuru, a process I normally regarded as a particularly noisome hair shirt, for it always went on far into a precious Saturday afternoon. Early that morning there were some Greater Flamingos out on the rocky islands a kilometre and a half from their October breeding ground. They were still there that evening as I passed on my way back. Easter was coming, three full glorious days in which I could really hope for success. By 19 April, when I returned to Elmenteita, 1,590 odd pairs had laid, within four or five days of their first emergence on the islands. Others were coming out on the islands used in October, evidently seeking to lay there too. The second chance, so often denied by Nature, was now to be offered, with both hands as it were. I set out to seize both hands and grasp them firmly if I could.

The first thing was to find out whether my floating hide would work. No such thing had ever been used before to my knowledge, and it was a toss-up whether a moving contraption without a visible man in it would scare them or not. I did not dare to try it till nearer the hatch, but busied myself making notes from a distance on the details of incubation behaviour and other things. Then, when they had been sitting for three weeks and seemed thoroughly settled, I tried it out. I chose to approach a small group of about 120 pairs carefully, feeling that even if I did cause them to panic and desert, the damage would not be critical in a colony that now numbered well over 6,000 pairs and was still growing.

I had to wade out through thigh-deep water overlying that same sucking impalpable mud I had felt beneath my feet when I swam in Lake Hannington. Now and again I stubbed my toes on submerged lava ridges I could not see – all my nails came off later. It was a windy day, and at twenty metres the necks of the sitting flamingos went up. Heart in mouth, I gripped the hide while they steadied again, then approached cautiously, step by step. My objective was a rocky shoal fifteen metres from them, where I could beach the hide and rest. As I approached it there was a loud 'clonk' as one of the galvanized drums hit an invisible boulder, and all the flamingos rose in panic and bolted into the water.

Disappointment clutched me by the throat again. However, I thought I would give them a little time. A few minutes later the first hesitant birds began to return. They withdrew almost at once; but it was a sign. When next they came they brought the others with them in a trumpeting body of red stilt legs, plumes all raised in territorial display, heads lowered to find their nests. Fifteen minutes after that first disturbance they were back on their eggs. I moved the hide and drifted past them. They sat tight. They had accepted this strange thing, and me inside it. It worked!

From then on for three months I was busy every week-end. I left the hide there, hidden in long grass, and camped each Saturday night in acacia woodland not far away. Day after day, in my cloak of invisibility, I worked my hide up to any chosen island where things were happening, beached it there, and then spent many hours watching, taking notes, filming, and photographing. Sadly, I lacked a camera with which to take colour pictures, and in those days could not afford one. However, my film is adequate testimony that I was so close to the flamingos that I could have reached out and touched them. The heads of the great males, standing one and a half metres (five feet) or more, were above the top of the hide and I saw them arched against the blue sky.

Once they had accepted the hide they paid absolutely no attention to it at all. Other people have since made other and possibly better floating hides, but it was my fortune first to study a colony of Greater Flamingos at point-blank range, able to move about among them at will, pick my island, observe any facet of behaviour I chose. Faced with such a surfeit of new facts, it is impossible to grasp it all at once, but I recorded most of it, with an opportunity no one had ever had before, for all previous studies had been done from fixed hides with rather shy flamingos. At Elmenteita I could have played recordings of a brass band and no flamingo would have paid any attention, for the sound would have been drowned in the trumpeting clamour from a thousand throats.

At length, about June, I was able to draw breath. Apart from rare week-ends the flamingos had taken up all my energy and spare time. I had notebooks packed with details hitherto unseen, scores of good photographs, and thousands of metres of film as a permanent record of the event. I even knew that my exposures had been right; and if they had not been, I could have returned and taken the same shots again. However, I took it on Kodachrome II, truer colour-wise, but at first unsaleable to BBC television, or so they said. Later they did make use of it, in my first 'Look' programme on flamingos.

Many times since I have done the same thing at other colonies and I have grown a little blasé about it. I do not see anything much that is new any more. All the same, each time I use my hide to approach a colony of Greater Flamingos I am thrilled anew by the wonder of the sight. There they all are, individually grotesque, yet somehow aristocratic. They perform antics that, objectively, are ludicrous in the extreme but, in the mass, form the greatest ornithological feast of beauty that anyone ever saw. It is unbeatable; even the crowds of Lesser Flamingos do not have the same overpowering effect of transcendent beauty, pulsing with excited life.

The only drawback was that they were not Lesser, but Greater Flamingos – birds whose breeding habits had already been quite well studied elsewhere. With the Lessers I got no further. In a few hours' flying of an evening I could assure myself that they were not breeding on Lake Natron. And, supposing they did, I would somehow have to find a way out on to those awful mud-flats again, which had so nearly cost me my life only two years before. I could learn a certain amount from the air, but obviously not the sort of detail that I could gather so easily from my floating hide on Elmenteita. Some day I would have to face those mud-flats again – or so it then seemed.

There were heavy rains in 1957 and Lake Natron became a pool of shallow claret-coloured water. I flew over it one fine June evening, and only one small patch of the crystalline mud-flat peeped out of that glass-calm, deadly red pool. It was an evening of sublime beauty, with all the great mountains round the lake mirrored in its waters. However, having satisfied myself that no flamingos were breeding, I took care to fly high enough to glide to either shore at need, for I knew that although the soda-flats looked beautiful covered with a film of rosy liquid, should I be forced to land in it the plane would cartwheel and, even if I survived the impact, I would probably not escape alive.

A huge concentration of Lesser Flamingos, more than a million strong, now developed at Lake Nakuru. I have never seen this now famous spectacle better – and in those days I had it almost to myself, for Lake Nakuru was not then a National Park. At the south end of the lake there was a stretch of shoreline on Nderit estate, totally undisturbed, and here I camped from time to time, watched the flamingos displaying, and vaguely hoped – with no good reason – that they would breed. Needless to say, they did not, though they did make a few mud-mound nests that were washed away by steadily rising water. They displayed in thousands with unabated vigour for months on end and between 8 and 18 September suddenly left the lake in force. At least a hundred thousand must have flown out nightly. I flew over Lake Natron between these dates and there was no doubt they had gone there, for the numbers had increased hugely. Perversely, they had mostly lost the power of flight. Flamingos were supposed to do this *after* breeding, as do ducks. So I supposed that, frustrated by rising water from breeding at Lake Nakuru, they had reached a stage in their physiological cycle when moult they must – and did – and that they would not now breed that year.

However, if I knew anything about flamingos by now it was that they abided by no rules. No firm predictions could be made as a result of what seemed to be reasoned observations. So I flew again over Lake Natron on 19 October, half-expecting to find nothing. And there, at the toe of the Gelai mud-flat, where we had seen that first group of young ones in 1955, was a huge dark-red patch. It could be only one thing, and was – an enormous breeding colony of Lesser Flamingos. My heart sank when I saw where they were breeding, among the crystalline slush of the mud-flat where the water receded, a good twelve kilometres (eight miles) out from the shore, far beyond where I had nearly died. Caring nothing for the comfort of my unfortunate passenger (who had said he would not be airsick, but was vomiting steadily into his shirt-tails and had, moreover, eaten a substantial meal of steak and onions), I circled about until I had gathered what details I could, then flew back to Nairobi. That year was evidently destined to be an *annus mirabilis* for flamingos. The question now was whether I could do anything about it.

As we flew back it seemed that I might, after all, reach a colony, for at the north end of the lake there was a smaller group of nests. They seemed only about six kilometres (four miles) out, and the going looked good, over quite hard brown mud. They had apparently just laid and were sitting steadily. At least I could try – and with care I would survive.

A week later I made an attempt. I camped near a swamp at the mouth of the Uaso Nyiro River, where the myriad mosquitoes bit like dogs all night and I could scarcely sleep. I had a new set of mud pattens made by Myles North and this time proposed to take with me the faithful Njeru Kicho, gum-booted, to help carry gear. Knowing more of what we had to face I had made better preparations this time and was ready to turn back at once if things looked at all dangerous.

We set out just as the sky in the east began to lighten. The colony was invisible, far out in the lake in the mirage, but Njeru, to his eternal credit, trusted in me. I told him he would first see the birds writhing in the mirage in a certain direction. In due course they did appear where I said they would, and from then on his confidence never wavered. I had again misjudged the distance: the colony was at least six kilometres out on the mud-flat. It was not crystalline but sticky and glutinous, and damned hot. My mud pattens again proved completely useless – they stuck to the surface like glue, and I was literally rooted to the spot. Njeru was convulsed with laughter by my struggles, but we went on, and reached the colony at about nine o'clock.

It was already grilling hot out there on the mud-flat and there was nowhere dry to sit or put anything down, except one little island of crystalline soda, where Njeru spent most of the day. I erected my canvas hide that I had first used at Lake Hannington, but to my surprise the flamingos would not accept it. So I did not take any close-up film or photographs but had to withdraw and make do with some rather indifferent middle-distance shots. However, we did make some sample counts of nests. We observed details of the incubation behaviour and in fact did all we could as far as the conditions allowed.

The scorching heat, at first tempered by a breeze, bore down on us at midday like some monstrous overpowering weight. In the heat of the day we were literally being slowly cooked alive. I re-pitched the hide a little nearer the flamingos, watched them accept it, and then started back, knowing that a week or so later I would have to repeat this awful struggle. We had both tanked up to capacity beforehand ('vomiting point', as Lawrence of Arabia called it) and had each drunk about a four and a half litres (a gallon) since, so we were not too badly dehydrated. But we were exhausted, and the trek back through mud that felt even more gluey than on the way out seemed endless. We encouraged each other by saying 'not so far now' or 'more than half-way now', and the bulk of the mountain Shombole increasingly towered above us. At length, when the worst sting had gone out of the sun, we found ourselves walking on hard caked mud, along the line of our outward track. It was over – until next week.

Njeru sat down on an isolated rock. We had come to no harm, but were weak with fatigue and dehydrated anew. With characteristic wit he observed: 'If a cup was given for bird-watching we ought to have won it today.' He deserved it more than I did, for he had come with me into a frightful place he knew had nearly killed me once before, trusting in me. He must be the only African who has ever stood on both the glaciers of Mount Kenya and the burning soda-flats of Lake Natron.

I dreaded the inevitable return a week later; but I was spared. Torrents of rain washed out the whole colony, which I checked from the air. I could see no sign at all, in the wave-lashed muddy water, of my hide which had begun its career on the shores of Lake Hannington and ended it here beside this colony of Lesser Flamingos.

Almost with relief, I turned to extracting what detail I could of the breeding of Lesser Flamingos by regular aerial surveys. I estimated the incubation period at 28 days and the fledging period at 70 to 75 days, while the development of the chicks was clearly similar to that of the Greater Flamingo. And I felt that if these confounded birds could not some day breed in a more accessible spot, the adventure was ended in this quest as far as I was concerned. I felt that never again could I face walking over those terrible scorching soda-flats; and to this day whenever I fly over Lake Natron I feel a shrinking in the pit of my stomach. To me it seems like a blotch of some ghastly disease on the earth's surface, even if reason tells me that it is only an expanse of crystalline soda and mud, the natural product of the environment, not meant to be malignant.

Later, my wish was gratified. In 1962, with their characteristic unpredictability, Lesser Flamingos bred at Lake Magadi for the first time in living memory, no doubt because Lake Natron was full of water and they could not breed there. I was hard worked, enduring one of my periodical stints as Acting Director of Agriculture which, among other things, forced me to waste time in innumerable board meetings. However, Alan Root, who camped at Magadi for months while making a film of the flamingos for Anglia Television, was able to make much more detailed observations. Between us, we dotted most of the i's and crossed most of the t's that I had not been able to detail from the air. It was specially satisfying to find that crude estimates of numbers made from the air were not as inaccurate as might be thought. A twenty-minute flight by me, and scorching days of pacing and sampling on the soda-flats, resulted in two estimates of total numbers within 10 per cent of one another. There were really no surprises. The Flamingos behaved much as they should have done, though the fledging period was longer than previous estimates, perhaps because the adults had to fly to Lake Natron for food, and returned only at night. They did not moult to flightlessness – it would evidently have been inconvenient. A succession of tourists in ordinary cars watched this spectacle, which had cost me such bitter effort to view for the first time. By the end of 1962 we knew most of what I had set out to learn ten years earlier, and which had so nearly killed me in learning it. We ringed over 8,000 young but, as mentioned earlier, learned very little from this great effort.

We now know that Lesser Flamingos lay eggs on mud-mound nests, just like any other flamingo. But their colonies in the middle of Lake Natron are invisible to people on the shore because of mirage, and the young always trek from these colonies to the same places, at the toe of the Gelai mud-flat and in a small central lagoon, where they form into vast khaki-coloured herds, fed there by their parents until they are almost able to fly. Then they suddenly appear along the shore, able to look after themselves, in hundreds of thousands. To Kipkwe of Maji Moto and to others like him, it must seem that they are suddenly spawned, full winged, from the interior of Lake Natron, as if the product of that awful place itself, and not of an egg like any other bird.

Greater Flamingos have again bred repeatedly on Lake Elmenteita; and everyone who has watched the 'Look' programme on BBC television has seen my floating hide – which has been cribbed by every other television company too. It is a good joke to see those silly disembodied feet paddling in the mud beneath the swaying contraption. That does not alter the fact that it was a brilliant idea that worked, like Cherry Kearton's stuffed cow. It works to this day; and nowadays we do not worry if it clonks on a rock. They will be back!

No bird or animal that I have ever studied has forced me to such extremes of effort and suffering. And none has ever given me such a reward of beauty and excitement, when discovery flooded swiftly upon discovery, in the presence of thousands of delicate pink beings, unlikely as the proverbial borogove of Lewis Carroll, and which are, nevertheless, reared in the awful heat and unrelenting glare of one of the world's most dreadful places.

There is still much to learn about flamingos; but I gave up the study after ten years, for I had picked all the easy meat off the bone, and spent enough of my own money doing the research, repaid by books and television films in cash, and in satisfaction a hundred times over. Someone may be able to rationalize them some day, and find out why they do what they do. He will certainly get a Ph.D. out of it – today's scientist's union card. However, at least one attempt to make sense out of it has already failed, and the scientist concerned, Dr. Eckehard Vareschi (who deservedly got his Ph.D. regardless), has had the grace to admit it. In a revised version of my book *The Mystery of the Flamingos* I bet anyone who tried a bottle of champagne that, at the end of his stint, he would have a mass of unintelligible data, which he could put into a computer, which would not be able to make sense of it either. I have Dr. Vareschi's bottle in my wine cupboard; and my wife and I will toast him and his on our twentieth wedding anniversary.

No one can ever do it again as I did, almost alone, without funds, and just because those little red water nymphs were still the greatest local mystery in the bird world. Personally, I hope that no one ever will fully rationalize flamingos, and that they will remain the supremely beautiful, elusive, opportunistic, unpredictable beings I like to think they are.

D. TISCHLER

7. The Honey Badger

For its size, I suppose the Honey Badger or Ratel must be easily the strongest animal in Africa. Its only likely competitor, weight for weight, might be the northern Wolverine. From films I have seen a Wolverine is about the same shape as a Honey Badger, and moves with the same clumsy lolloping gait. However, their habits and haunts are very different. The Wolverine is a carnivore of the northern coniferous woods capable of killing deer. Honey Badgers inhabit tropical and sometimes subtropical Africa and Asia, and live mainly on the grubs of bees and wasps.

Honey Badgers are not real badgers, though they vaguely resemble them and belong to the same family, the Mustelidae, which also includes weasels, otters, and Stoats, though not the rather similar-shaped mongooses. Whenever I see any member of this family I am glad they do not grow any bigger. Just think of a Stoat the size of a leopard, almost a hundred times as heavy as it is, and capable, by logical projection, of killing a young elephant. A Honey Badger as big as a Tiger just does not bear thinking of at all. I often wonder why such powerful effective predators, certainly capable of killing animals very much heavier than themselves, lithe, active, swift, and deadly, did not grow any bigger. Perhaps the answer is that they did not have to – they succeeded well enough as they were. Or maybe the wild dogs and cats, swifter and perhaps more adaptable, which take over as predators at about the upper size limit of the weasel family, had the superior agility and speed to kill bigger animals in the open, and so could compete more effectively. Yet a Stoat or a weasel 3 metres (10 feet) long would be a terrible creature.

An adult male Honey Badger is about a metre (3 feet) long, massive, heavy-bodied, and low-slung on short, rather bandy legs. It only vaguely resembles a true badger in its grey upper fur and black belly, but lacks the pied facial markings of the brock. Living very largely on the product of bees' and wasps' nests, it has an enormously thick loose skin impervious to stings. If gripped by a dog it is said to be able to slew itself round inside its pelt and grab its attacker. I do not suppose for a moment that in fair fight, dog to badger and weight for weight, the outcome would be in doubt, even if the dog were my formidable 32-kilogramme (70-pound) bull-terrier Patchy. I have often been glad to find, after a hard swift run through the bush, that my bull-terriers had seized nothing worse than a Wart-hog and not a Honey Badger. Gun or no – and sometimes I did not even have a stick – I would not have relished the prospect of saving my brave but utterly reckless dogs from an angry Honey Badger.

Honey Badgers apparently fear nothing, and give way to no one. They go their own way, and most other animals get out of it – though I doubt if an elephant would. If thwarted or angered they are said to attack and kill animals many times their own size, just because they are in the way. Their method is said to be special – they bite at the inguinal region of the victim, between the hind legs, until it dies from loss of blood. I do not know that anyone has ever seen this happen, but there are credible accounts of animals such as Wildebeest found dead with no other cause than mysterious wounds in the inguinal region. African hunters in Nigeria certainly believed implicitly that if they met a Honey Badger it was best to leave it alone, for fear it would go for that vital spot. They were in as good a position to know as anyone. If I met a Honey Badger in the bush, and it came towards me, I would certainly give it a wide berth; and if one suddenly came up to me like that British Badger in an Oxfordshire wood I would take to my heels, and no mistake. I think they are not very fast, and could be outrun.

African honey-hunters know about Honey Badgers as they themselves are in direct competition for the honey and grubs upon which the Badgers feed. They hang their hives in trees to keep them out of the way of the Honey Badgers, and contrive special slings too thin for them to climb down. Anyone who tries to keep bees in ordinary hives in Africa soon finds out that there are Honey Badgers about. The hives are despoiled, smashed open, and the honey and grubs eaten. Honey Badgers can climb, but not very well, so if a hive is placed high enough in a big smooth-barked tree, or slung on a wire or a thin stick, hooked at the top, it is reasonably safe. A specially good bee tree is sometimes protected by a circle of sheet metal – this defeats the Badger's grasping claws. The claws are not sharp and retractile like a cat's but are powerful, for bursting into hollow trees, and the animal is heavy and rather clumsy.

The lives of the Badger, the honey-men, and a small bird called the Honey-guide are inextricably interwoven in a fascinating web which, as much as any single feature of recent prehistory, must have altered the ecology of Africa, through fire. The bird, the Greater Honey-guide (*Indicator indicator*, the guider who guides), leads both badgers and honey-men to bees' nests. The relationship is symbiotic, that is, both benefit, though not at the same time. The bird does not need to eat the honey or the grubs – it is after the beeswax, which the others concerned do not want, or not so much. Honey-guides have special enzymes in their digestive systems which enable them to digest wax, otherwise one of the most indigestible of all carbohydrates.

There are many species of honey-guides and honey-birds, of which two are Asian and the rest African. Relatives of barbets and woodpeckers, many of them parasitize their near cousins and some other birds. Since they normally lay their eggs inside deep barbets' or woodpeckers' holes the newly hatched young bird, unlike a baby cuckoo, cannot just lift the unhatched eggs of the host on its back and throw them out of the nest. The difficulty is overcome by a truly astonishing development – a pair of sharp, hard, pincer-like hooks with which the blind naked honey-guide nestling bites and kills its nest-mates, or punctures the other eggs so that they do not hatch. Cuckoos seem malevolent enough in their blind attempts to eject their nest-mates, but one feels that Satan had a hand in the creation of the honey-guide's hooks.

Honey-guides are among the most extraordinary and least known of African birds. There is still argument as to how many species there are and, although we presume that they are all parasitic, the eggs and young of at least six species have never been seen. One makes a strange trumpeting sort of sound high above dense forests, by the air passing through its specialized tail-feathers, just as a snipe drums. Of the many species, only one actually guides regularly, and that not everywhere. The Black-throated or Greater Honey-guide is the only real guider and in parts of Kenya it does not guide, though males are heard calling from treetops in season, so one knows it is there. This has been explained on the grounds that, with increasing civilization, a knowledge of Shakespeare and New Maths has superseded the need to know where to go for honey. However, this explanation does not satisfy me, because during the hundreds of hours I spent in the company of keen Wambere honey-men in uninhabited bush rich in wildlife, I was seldom guided. They certainly wanted honey and knew of the guiding habit, but just said local birds did not do it. They would certainly have followed the Honey-guide if it had done its work in Mbere.

Elsewhere it is very different. In the savannas of Nigeria and western Ethiopia, and in many other areas, the Greater Honey-guide almost invariably tries to guide any human being it meets to honey. Human beings also seek the aid of the bird and try to attract its attention. The bird guides by emitting an insistent, rattling call, said to resemble a person shaking a box of matches. In Nigeria, alert wild animals constantly hunted by Africans associated the call of the Honey-guide with the presence of human beings, and bolted at once. One day I followed a solitary bull buffalo from before dawn to near dusk; twice crept up on him as he lay in a thicket of thorn; and lost him each time because a blasted Honey-guide found me, rattled, and warned him to leave without waiting to see why.

Surma honey-hunters in the Ethiopian Omo Valley spread out over a wide area, tap trees, and whistle to attract the bird. Given a broad enough front, one man will soon locate a bird, and all can then follow his call to the wild bees' nest. This may be in an inaccessible and impenetrable crevice in a crag, in which case the honey-men must leave and try elsewhere. Then the bird follows desperately, flying round in front and chattering, as if to say 'What's the matter with you? There's a perfectly good bees' nest where I showed you.' More often, probably, the nest is in a hollow tree, and an axe soon exposes the comb and the grubs. African honey-men will not depart without leaving some comb for the Honey-guide, for they believe that if they do not, next time it will guide them to a venomous snake or a sleeping leopard.

How men learned to use Honey-guides, or vice versa, is a moot point. Some say they learned it from the Honey Badger, which the Honey-guide also leads to bees' nests. Naturally, few people have ever seen this happen, for one sees a Honey Badger itself seldom enough in the wild. However, Nigerian hunters I know used to grunt deeply, like a Ratel, when calling to the bird. For years Alan Root had a male Honey Badger confined in an enclosure; and almost as soon as he acquired it, a Honey-guide appeared and chattered – in an area where most Honey-guides seldom guide men. Unfortunately, he never recorded this on film, because the Badger died of distemper before he could bring bird and beast together in an area where Honey-guides regularly perform to order and where he could arrange the action filming to his satisfaction. As a matter of fact I was once guided within a few kilometres of his house, but though I told him about it the Badger was then dead.

It is also said that Honey-guides may attempt to guide baboons; and if so, what more likely than that they would guide primitive, tool-using hominids of a million years ago, scavenging for food through the bush in troops? Such primitive men, our ancestors, had stone axes which would serve far better than the delicate fingers of a baboon to break open trees. These hominids may have seen the bird guiding the Badger; and when we ourselves meet a Honey-guide in the bush, and are guided by it, we may be touching an ancestral memory that goes back into the far mists of our origins.

Certainly, it is salutary to think how vast has been the effect of the interwoven lives of men and birds. Savanna or bush fires are sometimes started naturally, by lightning or some such agency; but nearly all of the

13,000,000 square kilometres (5,000,000 square miles) of Africa that go up in smoke each year are burned by man. These fires that start mysteriously, in the middle of uninhabited country, in National Parks and elsewhere are started nine times out of ten by secretive honey-men. They need the fire to make smoke to get at the bees' comb without being stung. They make it in the age-old way, by twirling a hardwood stick in a notch in another piece of wood – if you do not carry something really modern like a Nigerian hunter's flint and steel. The fact that the ground is burned is also useful to bare-footed men as this exposes Puff-adders and other such dangers of their daily lives, and makes cross-country walking much easier.

Men have been able to use fire for at least 350,000 years, perhaps longer; in that time how many uncountable fires have been started at wild bees' nests, found with the aid of the Honey-guide? So many, certainly, that the wooded face of Africa has often been turned into tall grass savanna or grass plains over a vast area and is still being changed or maintained in this way. I have seen it done, often, in the sixties and seventies, just as it was maybe half a million years ago.

How and when men learned these ways is perhaps immaterial, but they may have learned them through watching the bird and the Honey Badger working together. There are few authentic cases of reputable observers seeing a Honey-guide lead a Badger, and few people have kept the Badgers captive, as they are such intractable beasts. One seldom meets a Honey Badger by day, and when one does there is an indefinable something about the animal that prompts me, at least, to keep my distance. It carries an ineffable air of menace that inspires caution.

In long grass the Badger is soon lost to view, and I certainly would not want to follow too closely, even if I saw it being guided by a Honey-guide, unless I were equipped with, say, a cricketer's box. People do carry odd, possibly useful things such as snake-bite outfits about in the bush with them. However if, like the White Knight, one carried everything one might conceivably want, one would soon be overburdened. I never yet heard of a scientist who carried a cricketer's box about with him on the off chance that he might meet a Honey Badger being guided by a Honey-guide and wish to take notes about the detail of the subsequent attack on his inguinal region. One must draw the line of reason somewhere.

Alan Root's Honey Badger was an impressive animal. It was as big as a bull-terrier and no doubt much heavier and stronger and it could and did break out of very strong enclosures several times – only to come back of its own accord, when it found food not so easy to come by. Alan acquired it from Armand Denis, and it was believed to be incurably savage. However, Alan concluded that the grunting animal running up and down behind bars only wanted to be played with, and disregarding the risks to his manhood, as he disregards so many others, leapt into the enclosure and played with it.

Although there is undoubtedly a fascinating interconnection between the lives of men, badgers, and birds, there is still comparatively little undisputed fact to underpin all the imagination and conjecture. I cannot count the thousands of hours I have spent walking through African savannas, where Honey-guides abounded, and Honey Badgers were certainly as common as ever they are. Sometimes I was alone, but more often with hunters or honey-men, who would at once have noted any interaction between the beast and the bird. I have been abroad at all times of the day and at all seasons, when the grass was burned and the ground black and grey with ash, or when the new spears made a carpet of emerald green after rain and everything was bursting with new life. Then again, when the grass was tall and over our heads, and I could see why anyone would want to set it on fire, just to make walking through it easier.

Thus I have scoured the bush over thousands of square kilometres and many years, hunting or looking for eagles' nests. Yet, for many years, I could count on the fingers of two hands, if not only one, the number of times I had caught even a brief glimpse of a Honey Badger. Even then, I never had one in view for more than a few minutes, as it moved purposefully along, paying no attention to me, while I usually had some other objective in view and so let it go its way. Also, there is that air of menace, which discourages too close an inspection. It is different from a car where, in National Parks, people quite often see Honey Badgers and take their photographs.

Then suddenly, in 1964, the moment of revelation came. Like most such moments it was unsought and unexpected. I was in the Kalahari Gemsbok National Park, looking at the area and acquiring material and photographs for my book on the natural history of Africa. Personally, after the scenically magnificent National Parks of East Africa, I find the Kalahari Gemsbok Park rather dull, for really all one can do is to drive along a track down or up the valleys of the Auob and Nossob rivers. They only flow in very heavy rain, but their subterranean moisture supports large acacia trees. These are great places for certain birds of prey, especially Martial Eagles and Bateleurs, and many a well-known and much-published photograph has come from here. There are also the huge, haystack-like nests of Social Weavers, the Gemsbok themselves, plenty of Springbok and, if one is lucky, a spectacular Lion. But I did not see any Lions; and it was very hot and dry.

In two days I had seen all I wanted of the area, but nothing unusual, and on the afternoon of the second day was driving back up one of the river beds to camp. Suddenly I saw an animal lolloping over the red sand dunes about two hundred metres away on my left, among thin scrub. It was a Honey Badger, a big one, right in the open in the middle of the day, apparently oblivious of my presence and caring nothing for a moving vehicle. It loped along rather clumsily, and it was then that the likeness to the Wolverine I had seen in films struck me so clearly. It was running over the dunes parallel to the track, and though I did not expect to keep it in view for long I followed, the Land Rover chugging along of its own accord in low ratio.

I expected the Badger to return into the dunes from where it had come but it did not. Instead, it kept on parallel to my course, keeping about the same distance away, usually too far for me to photograph or film it, but near enough for me to watch it closely through binoculars. Although it could doubtless hear the engine of the car it paid absolutely no attention, and seemed to be behaving in a completely natural manner. When I restarted the motor after a temporary halt it looked in my direction, but did not take fright; no doubt it had been watched before, in the same sort of way. In all, I watched it continuously for almost five hours. Occasionally it was out of view behind a dune or a clump of bush, but always it reappeared, lolloping along parallel to me as before.

There are no African hives in the Kalahari Gemsbok Park, but there are plenty of hollow dead limbs where wild bees might harbour, so I had great hopes that I might actually watch it rip open a bees' nest. Alas, no opportune Honey guide appeared to guide it. Instead, it was itself followed for periods by a Pale Chanting Goshawk and a Spotted Eagle Owl. Both these predatory birds, one diurnal, the other nocturnal, left their perches as the animal passed, and followed from tree to tree for several hundred metres. Here was another probable symbiosis, of which I had heard from the park warden and other visitors. The birds were after rats or lizards that might be disturbed by the hunting Badger.

This particular Badger spent a lot of time digging; and no doubt the birds hoped to catch any rats which it disturbed from their burrows. If so, they were disappointed, for as far as I could see the Badger got nothing while they were near. Although he – I assumed from its size that it was a male – dug assiduously, even disappearing underground, I never saw him catch anything. However, he could easily have found a nest of young gerbils underground and eaten them there without reappearing. Despite the express prohibition against leaving the track and driving over the dunes issued by the parks authorities I would have tried it if I had not been suffering from a slipped disc, which made it necessary to wear a steel support, and made digging a Land Rover out of deep sand a daunting prospect. So I just kept moving along beside him, hoping he would come closer. I could have got out of the car and walked up to him when he was underground, but I feared he would then take fright, and I confess I also thought of my inguinal region. Stupidly, I had forgotten my cricketer's box again.

The animal went on and on, doing much the same sort of thing. Sometimes he came close enough for me to photograph, but the light was poor and against me, the desert sky grey with dust haze. The results were poor, and he was often partly obscured by thin desert scrub when he was doing anything interesting. Then – needless to say as I was changing a film – he climbed a tree. It was not a high tree, and had no hollow limbs, but there was an abandoned Cape Rook's nest in it which he investigated. He went up easily enough, using his forefeet to grip the rough bark, then humping up with his hind feet, climbing in a series of clumsy jerks. Evidently he was not in the first class as a tree-climber – not even as good as a Lion – but could do it if he had to. He found nothing in the nest though he tore up the lining with his claws, and then backed down before I had finished changing the film. This he apparently found much more difficult. He was not more than three and a half metres (twelve feet) above ground and could not have hurt himself if he had just jumped on to the soft dunes below. However, he backed down slowly and carefully, casting glances over his shoulder, and moving only one foot at a time. It took him almost twice as long to descend as to ascend; and then he went on as before, loping over the dunes, digging at intervals.

He now made to cross the road at an angle, and this would put him in better light for photography. However, it was late afternoon and autumn in that part of the world, so that the low sun, obscured by desert haze, gave little enough light. Even at full aperture I could not give a short enough exposure for real confidence in the result, but I tried. I took a few photographs of him digging in burrows, and one at least turned out better than I had any right to hope.

Then he turned away from the road again but, to my regret, did not climb any more trees. He was accompanied by no more eagle owls or goshawks, and I could not be certain that he killed anything in the nine and a half kilometres (six miles) or so that he covered in four hours. Lolloping over that soft sand he must have used a good deal of energy, but I suppose that even a few nests of gerbil youngsters would have kept him going.

At about 5.30 he seemed about to disappear in the desert to the right but then turned back and made to cross the road at an acute angle again. The setting sun was now right behind me, and I thought that if I could position the Land Rover directly in his path I had a good chance of a unique picture. I positioned myself perfectly without disturbing him, and waited, camera aimed. He came loping down the side of a dune among low shrubs, and stopped when he saw the car in his path. He was too close – I could not get him in focus!

He regarded the Land Rover with what Kai Lung calls 'a two-edged manner'. Clearly he thought it had no right to be there at all, straight across his chosen path. There was no question but that his whole pose was one of menace. A Land Rover, strictly speaking, lacks an inguinal region, but I half-expected him to hurl himself at and blunt his teeth on the differential. Instead he swerved a little from his chosen course, passed across the road not far from the front wheels, and with a single angry backward glance went on with what he had been doing. I reckoned I would see no more, and I had to be back in camp before dark, so I left him to it.

It seemed yet another of those almost miraculous, prolonged encounters with a little-known animal, previously glimpsed for a brief moment less than ten times, in a lifetime spent in suitable habitat. Certainly, I had watched him for more than a hundred times as long as I had ever watched a Ratel before. On reflection, however, I fancied that his behaviour might not be entirely typical. He was living in an extreme environment in that sandy desert, where the scarcity of suitable food may have forced him to spend more time than usual foraging. I have heard of other people who have watched Kalahari Honey Badgers out in the open for quite long periods. They have also been seen in the open Serengeti plains, digging for rats just as my animal did. The Ratel is primarily an animal of the savanna, like the Aardvark. At the fringes of the environment tolerable to its existence it would probably be forced to adopt ways of life unnecessary in well-wooded savanna, where half an hour's foraging would result in a willing Honey-guide, and a gorge of bees' larvae and honey. All the same, I am glad I was there and saw it all so closely, even if I did not film and photograph it as well as I might have done.

Maybe, some day, I shall meet another Honey Badger in more normal conditions and see him guided by a Honey-guide – which will also guide me to what we all want. If so, I shall have to depend on a stout club for protection, if I can find one. If the worst comes to the worst, I now know that I can climb any ordinary tree a good deal faster than he can and that I should have a good chance of escaping his menace among the higher branches.

8. The Mountain Nyala

The Mountain Nyala, *Tragelaphus buxtoni*, was the last of the major African ungulates to be discovered. It was first collected by Major Ivor Buxton in the Arussi Mountains of Ethiopia in 1908 and scientifically described in 1910, six years later than Meinertzhagen's Giant Forest Hog, and a decade or so after the Okapi. Only half a century later, in the late fifties, its continued existence in any numbers was in doubt. One authority, T. Donald Carter, who had collected a group for the Chicago Museum, opined in 1961 that there could now be only a few hundreds left. The pace of wildlife destruction in Africa generally and of habitat destruction in the mountains of Ethiopia since the Second World War made this only too probable.

I first read of the Mountain Nyala in the old Badminton Library series in a chapter on Ethiopia by H. C. Maydon. Even then I was deeply intrigued by that mysterious country, the largest mountain massif in the whole of Africa, which had preserved many of its mysteries intact well into the twentieth century, and for that matter still does today. Maydon was a big-game hunter, as in those days was I. He wrote of the African bush and hunting with a magic touch that no one has since bettered, and he endeared himself to me by quoting a verse of my father's (H.B.), published in *Punch* many years back:

> Ye who are sib to the jungle, and know it and hold it good
> Praise ye the ways of Nimrod, a fellow who understood.

Far away in Nigeria it seemed to me then very unlikely that I should ever tread in Maydon's footsteps, or see the intriguing beasts he had gone to Ethiopia to hunt; but I did. By then, though I had lost any desire to hunt them.

It cannot often fall to the lot of a scientist, wishing to study a beast whose survival is in doubt, to be able to correspond with its discoverer. However, I was able to write later to Major Buxton, and although his reply to me was brief, it was illuminating, for it showed that the essential habits of the Mountain Nyala had not changed from the day it was first seen to the present. There is no essential reason why an animal's habits should change. However, the habitat had certainly been changed very drastically, so it was worth finding out from the man who first saw the beast in its natural haunts and collected it for science. He thought it was some sort of kudu, and did not realize it was a completely new species until later.

What little was known of the Mountain Nyala in 1963 indicated that it lived in the high heath zones and higher forests of the Arussi and Bale Mountains of Ethiopia, a wild stretch of thinly inhabited highlands with a vile climate. Think of the bogs of Sutherland covered with head- or waist-high heather, sheeted daily in heavy icy rain, with frost almost every night of the year, and you have it. Yet that country sometimes too has days on end of sublime clear sunshine, when one feels one is on the very roof of Africa, in a place where very few other than the local inhabitants have ever trod. The sun grills and flays one's tender white skin by day, and the nights are piercingly cold, adorned with the myriad twinkling stars of equatorial heights. There are still glorious unspoiled forests of cedar and *Podocarpus*, with open grassy glades; or queer wet forests of Hagenia and giant St. John's Wort 15 metres (50 feet) tall, carpeted in season with flowering red-hot-poker lilies (*Kniphofia*). Crystal-clear streams drain these lovely forests into the Uebi Shebeli and Ganale Doria rivers, and picturesque Galla tribesmen live there with their livestock, still little touched by civilization – though gumboots and umbrellas are now *de rigueur*. I had learned a little about this country from Maydon, but I longed to see it for myself.

In 1963 the chance came. I escaped from Government Service, and began travelling all over Africa, gathering material for my book on the continent. Since there were several endemic, possibly endangered, mammals in Ethiopia, the Walia Ibex of Semien and the Mountain Nyala among them, I obtained a small grant from the World Wildlife Fund to cover expenses, and in October set off into the blue. I went first to Semien, the most magnificent country I have seen anywhere, not excluding Arizona's Grand Canyon; and then to the Bale Mountains, to look for the Mountain Nyala. Later I wrote a book (*Ethiopian Episode*) about it all, and with what I made from that and my small grant the whole three-month trip cost me nothing and brought in a small profit. That was not the object, but it was welcome to one who in his mid-forties had to seek another way of life.

I went to Bale with an American called Bill Corcutt, of whom I still think with affection, though we have lost touch, as the only thing on which we did not agree was politics. For some reason he had picked on Ethiopia as a place to spend a vacation and he came up to me in the Ras Hotel in Addis Ababa and asked me if I could advise him where he could get some good tramping. Cutting myself to make sure the blood still flowed, for Americans in those days were not as keen on tramping as they are now, I said I knew of at least one such place, and was going there soon. Would he like to come? It was all settled over most of a bottle of whisky, which settles so many things amicably, and we set off for Bale on 16 November. I had been told in Addis Ababa by people who should have known that it would be dry in the Bale Mountains in November, and so took no tents other than my own small mountain tent. How wrong they were, for it turned out wet as wet can be.

We first climbed up into the mountains behind Dodola where Bill, who had the sensitive interior common to many Americans, learned something about Africa the hard way. Later in our more scurrilous moments my publishers (Country Life) thought of calling the book *The Men's Room at Dodola*, but they funked it in the end. Above Dodola we saw some Mountain Nyala, but all cows or young bulls, and none very close. My most memorable experience was finding a Lammergeier's nest at 2.15 of a wet afternoon and having photographs of the bird on the nest – still the best I have taken – by 3.30, using only a rough hide of heather branches perched on the very lip of a steep precipice.

During this week we had no tent for our camp followers and guides, who suffered badly out in the open for several nights. At the end of it they were justifiably rebellious, and after one all-day trek through cold soaking rain we found shelter in a Galla herdsman's hut. I shall never forget the perfect poise and grace with which the young wife welcomed these strange foreigners. She may never have seen a European before, but she did what honours she could with distinction.

We still had another week but just had to find somewhere dry in the high mountains for our people to sleep. So Bill and I climbed, alone, into the massif near a big mountain called Cotera, to look for Nyala, and to find a dry place to spend a week while we did further surveys. Bill had already shown that, apart from his delicate insides, he was virtually indestructible; and I can still see him in my mind's eye, plodding through the rain and the bogs, water dripping from his snout and hat-brim, but game to try anything.

On that day the weather was not too bad. We climbed to 3,500 metres (11,500 feet) over stony ridges, meeting no one once we had left the homesteads behind. We were looking for a cave or an abandoned hutment used as temporary accommodation in the dry season in which we could lodge our people. In due course we found some huts, in a lovely spot overlooking a splendid forested valley, loud with the voice of the stream in its depths. Here we sat on a rock terrace and watched a soaring Verreaux's Eagle, and a Semien Fox – another mammal unique to Ethiopia – ratting in a sunlit forest glade. Then we set off back, satisfied that we had found what we were looking for.

The first sight of an animal one wants very badly to see may not be very satisfactory, and so far we had seen no Nyala very close, nor any big bull. However, if one perseveres, the moment comes and one sees the beast to perfection, suddenly, unexpectedly, and with breath-taking impact. We were working along a rocky ridge, looking into a valley where the heath had been burned, and Bill paused on the lip of a rock ledge. Beneath him there was a patch of unburned heath, perhaps ten metres square and over a metre high. From it there burst a huge dark slaty Nyala bull, who had lain there like a rabbit till we were almost on top of him and he was certain he had been discovered. Running like buck ourselves, we were able to watch him gallop flat out along a bare hillside, jinking and twisting as if to avoid the expected hail of rifle bullets that may often have followed him, his splendid lyrate horns laid back on his withers, a superb beast in his natural setting. I had been right when I read Maydon – the Nyala was worth coming for.

We spent a week in our abandoned Galla homestead. Daily Bill and I crawled out of our small tent, to be almost instantly soaked with freezing cold drip from the tall heath, and into the hut for breakfast. It was not much of a place, some three and a half metres (twelve feet) across, round, made of bamboo and mud with a thatched roof, and with a thick layer of dry cattle dung on the floor. Yet it made the difference between a house and no house; and that is the vital one. Our people could sleep in there, dry and content, for they did not mind the rats and fleas – they had been used to these all their lives. Each day we explored a new area of mountain country, and every afternoon the clouds massed, soaking us to the skin, and deeper, with freezing rain before we reached home. I have never been so wet and cold anywhere in my life, even in Scotland.

Nothing levels the human race so much or so quickly as shared discomfort and misfortune. Daily we returned to the hut soaked and half-frozen, our skins like wet papier mâché. As soon as we were inside all of us, African and European alike, stripped naked without any sense of shame or embarrassment, and toasted ourselves round the smoky fire of dry heather stalks which the camp caretaker had made in the middle of the dungheap. We hung our clothes from the rafters to dry and did not care what we looked or smelled like. Bill and I sparingly shared nips from a bottle of whisky, while the others had their own form of potent spirits. Talk soon flowed as we thawed and dried and despite the acute discomfort we were all fit and happy.

Our African companions – for that is what they were – would have endured the suffocating smoke inside the hut. Bill and I could not, so we made a small hole in the thatch which allowed it to escape. This simple act nearly had fatal consequences. One morning, emerging from the hut after breakfast, I found that we were surrounded by a grim-looking ring of over a hundred spearmen and rifle-armed Galla. Since they did not immediately shoot me, or even aim at me, I hurriedly dragged out my terrified cook – my only interpreter – and walked boldly up to an elder who looked like a ringleader, feeling anything but brave. My cook was shaking and grey with fear, but the elder immediately burst into shouts of laughter. Slapping his thigh, he cried 'Faranjoch!' – just foreigners! – 'Why, we saw the smoke and thought you must be bandits!' It was a nasty moment; and to this day I think it was a lucky thing that I, and not an African, came out of the hut first. An African might have been shot on sight; and then they would have had no alternative but to blot the rest of us. We survived, and I determined to return, when I could, for a longer exploration of what I now knew to be a fascinating if lonely and uncomfortable mountain land.

In ten days, in only a small fraction of the total known range of the Nyala, we had seen or contacted 54 and found the tracks of at least 50 others. So it was barely possible to believe that there could only be a few hundred left alive. They were evidently about as common as such an animal ever was round our last camp. It therefore seemed that there was no special urgency to conserve a small dwindling remnant population, as so often is the case, and as I had found to be needed in Semien with the Walia Ibex. I could plan the next stage more thoroughly, make fuller preparations, and really study the status of the Nyala over a big area of country in some depth.

In 1965 I went to Ethiopia for three months on a UNESCO advisory mission to advise on conservation. I prepared a five-year wildlife conservation plan, never to this day effectively acted upon, which was to set up a basic network of National Parks and reserves to safeguard the future of endemic animals and habitats, attract tourists, and provide revenue for a budding Wildlife Conservation Department. It all seemed quite hopeful; but although some success has been achieved since then, it has not come to much, and such progress as was made has had to be abandoned following the revolution in 1974 and the subsequent political unrest.

In this plan, as a plum for myself, I set up a three-month research project on the Mountain Nyala. The object was to survey the whole known range of the animal, estimate its total numbers, and on the results base a National Park which would permanently safeguard an adequate population. I had an idea that this National Park would be in the Bale Mountains, for it was plain that the Arussi Mountains – where the first studies had been made – had now been largely destroyed by encroaching cultivation, forest destruction, and fire. The same old story repeated so often all over Africa, and still being repeated now when people ought to know better, and in fact do know better.

Working on the experience gained in 1963 I concluded that there was only one practical way to cope with the problem. That was to travel as light as possible, but with basic comforts, above all good tents and warm sleeping-bags. I would stay high up as long as I could, camping for several days at various sampling points in the Mountain Nyala's range, and send a horseman to the lowlands to buy basic supplies – bread and meat – in the village markets on the plains. I knew this plan would work, and that it would allow me necessary time for field-work in the haunts of the animal I sought. I tried two sources for finance, the World Wildlife Fund and the National Geographic Society, and to my surprise found them fighting for the honour. So I split the costs between them and carried out my expedition for a total of under £1,000 spread over three months, including air fares from Nairobi. Without my good friend Emil Urban, with whom I stayed in Addis Ababa, it would have cost a little more. It can be done quite cheaply, if you are prepared to live rough and on the basics – which include whisky.

I arrived at Addis Airport on 20 January 1966 with boxes of stores I had bought in Nairobi, and for which I expected to be fleeced by the Customs. Amongst other things I had an ice-axe, not because I expected to find glaciers in Bale, for there are none, but to clear lumpy tussocks of grass to make a flat bed to lie on in my tent. There was a lady Customs officer, and she looked at the axe somewhat apprehensively. 'What's that for?' she asked. Taking my cue expertly, I replied, 'That's for hitting people with.' She chalked my boxes and let me through without another word; and ever afterwards, when I passed through the Customs, she gave me a nice smile if we met.

This unexpected bonus from a bureaucratic system notorious for protracted procrastination was the last for some time. It was the Timkat holiday, and everything came to a standstill for several days. Then I went to Pelican Island – about which more later – and only managed to see the new head of the Wildlife Conservation Department after wasting the better part of a week of precious time. When I did see Major Gizaw Gedligiorghis, whose duty it was to give all possible aid to visiting scientists helping his Department, and who had offered to provide necessary letters of introduction to provincial officials, I found he had done nothing whatever to prepare the ground. Without such letters I knew I could get nowhere; and Major Gizaw had known for more than six months that I needed them.

I expressed myself pithily to him when alone in his car, and as a result we flew together to Goba, the capital of Bale Province, to see the

Provincial Governor. The place was in something of a turmoil, for Somali-inspired brigands were fighting Government troops in 1966, just as they were to do in 1977, but with rather less success. The Americans were supplying the Ethiopians with bombers, tanks, napalm, the lot; and the Somalis had not yet turned to the Russians for aid. Recently the whole absurd wasteful bloody political dance was being played out again over the same ground, with greater human loss, and changed partners. In case anyone feels like laying it all at the door of those dear old whipping boys, the colonialists, it was Emperor Menelik II who seized the Ogaden for Ethiopia.

When we arrived at Goba it seemed peaceful enough. There was no one on the airport, which is in the middle of the town, except the usual goats, donkeys, and swarms of humanity. From the aircraft as it drew up at the end of the runway I could see a few lackadaisical policemen and soldiers lurking resourcefully behind barbed wire. Major Gizaw was a noted big-game hunter, and an ex-Air Force pilot (removed from that post for breaking too many aeroplanes, I believe). Drawing his revolver, though not actually firing it, he sprinted across fifty metres of well-populated township to the Government offices, atop a small hill. I followed at leisure. Our French pilot asked, 'What's he running for?' When we got there Major Gizaw pointed out that this was an extremely dangerous area packed with *shifta* (bandits) and that we had taken our lives in our hands. I looked at the wild, misty Bale Mountains looming beyond the town, where I intended spending two months. I began to wonder whether I would ever get up into them with my little mule train, at best without an escort of yelling soldiery, and at worst whether I would preserve my manhood intact.

The Governor did not like the idea either. Mules and horses could not be bought or hired in these days, he said. I knew he must be lying, because mule and horse trains are the standard form of transport in the Ethiopian mountains, and I had seen plenty coming and going, just as I had in 1963. So I just said that I would arrive, with my gear, by the regular flight a few days later and that if he would not help me I would negotiate for mules myself, as I had before. I could see there was nothing for it but to arrive and make myself inconvenient until I got what I wanted. It is the only way of getting anything done in most offices in Ethiopia; the man you want cannot stay out for coffee *all* day.

As threatened, I arrived at Goba on 29 January, and managed to hire a Land Rover to transport my loads a few kilometres out of town to a camp beside the Tuguna stream. Here I at once sent my assistant and interpreter, Ato Mesfin Abebe, who had been loaned to me by the Wildlife Conservation Department, to negotiate for mules and horses while I with

Mick Prosser, a botanist from the University of Addis Ababa, set off to explore upstream. A few kilometres out of town we came on the edge of the cedar forests. Here in a cliff overhanging the stream I found the breeding site of a pair of Cape Eagle Owls, *Bubo capensis dilloni*. They were not breeding at the time, but I collected skulls, in about equal numbers, of common mole rats and of a fruit-eating bat. Unfortunately, the latter were lost, so we still do not know what sort of fruit-eating bat is caught by eagle owls at night at about 3,000 metres (10,000 feet), where the frosts can be sharp and there is little fruit.

When it became clear that I was there and that I would not go away the Governor, to his credit, decided to help. After talking with him in the morning, and hearing him instruct people to produce the necessary transport, Mick and I were able to climb to nearly 3,000 metres in the afternoon, through lovely open glades in the cedar forests, and make preliminary notes on the vegetative zones. This area was all heavily inhabited and grazed, and I did not expect to see any Nyala, but we saw baboons and bushbuck, and found a probable Lammergeier's nest – hopelessly inaccessible. It all seemed very peaceful; not only were we not attacked by bandits, but the nice Galla people of that part smiled and waved to us.

Next morning we packed up early, and – thank goodness – the horses arrived about eight, so we were actually on the move by eleven. It is best to turn one's back on the scene as horses are loaded in Ethiopia. The poor beasts are usually badly galled, sore from primitive saddlery, underfed and thin, but nevertheless remarkably tough. The yelling of orders and counter-orders that goes on would drive one mad if one could understand a word of what was said. I left it all to Mesfin Abebe. He had been sent out by the Department at short notice into what Major Gizaw knew perfectly well was a wild piece of cold mountain terrain, with no money and no tent, not even a blanket. Fortunately, I had anticipated this, and provided for him. He turned up trumps. One of our horsemen was a sixteen-year-old youth called Hussein. He was what the Irish would call 'a broth of a boy'; and he alone of all our men stayed the whole course, from Goba to Addis Ababa.

We made camp at 3,500 metres (11,500 feet), at the top edge of the forest, in a glade among the giant St. John's Wort, just where I wanted to be. We were in ideal Nyala habitat, and though we saw none that evening we found tracks, saw a melanistic Serval Cat, a Reedbuck, some Klipspringer and duiker, and made a collection of plants. As we returned to camp vast flocks of the endemic White-collared Pigeons, *Columba albitorques*, were flying up valley from the grain fields on the plain below to roost in cliffs at 3,600 metres (12,000 feet) or more. It was a fine night but cold, with a sharp frost; and we sat round the fire and yarned, scarcely able to believe that we had got so far on our journey within three days.

Next day we were up at dawn, and with guides who had slept at camp set off up a long ridge. We saw many duiker and Klipspringer, but no Nyala, and very few tracks. We climbed to the top of a peak some 4,120 metres (13,500 feet) high by about lunchtime; and there I sat on a rock and had my picture taken for the *National Geographic*. Unfortunately, I was wearing bush-green, intended to conceal me from a keen-eyed Nyala, and not a bright-red T-shirt; so although they have kept the picture I have never seen it in print. As I was sitting there I heard a sharp call, 'Kyaw', with a twangy timbre, and knew that I had in that moment extended the known range of the Common Chough, up to now only known in Ethiopia from Semien, by about 800 kilometres (500 miles) to the south-east.

On our way back to camp we did see and watch two Nyala cows, but there were not many about, and we were told by our guides that there might be more on the other side of the Tuguna Valley. So, two days later, we had to go back down, cross the gorge where we could, and climb up to another camp in the upper forest edge. It was a better, more sheltered, warmer camp than our first, and though I saw no Nyala the route I took was full of interesting birds and other things, in country that was all new to me.

Next day, 4 February, we were up early again. Led by two excellent guides, we went up the right bank of the Tuguna gorge, looking into it at intervals. At about 3,840 metres (12,600 feet), an hour or so after leaving camp, we saw a single Nyala cow watching four Semien Foxes, two adults and two grown cubs, which passed close enough for me to take a photograph. We then stalked the place where the cow had been and, peeking over a ledge, saw a magnificent bull and several cows right in the open. I was able to take some photographs before they bolted, and apart from rather dubious ones taken at long distance two days before, they were the first photographs of Mountain Nyala ever taken.

Action continued thick and fast. There had been twelve Nyala in the first herd, two of them big bulls. We saw three more cows and calves, which just melted away when I tried to stalk them. About lunchtime we climbed up on to a forbidding flattish plateau, at about 3,900 metres (13,000 feet), riven with deep gorges, and covered with short grass and *Alchemilla*. On small tarns here there were many migrant European ducks, Wigeon, Shoveller, and Pintail. Tawny Eagles and harriers were trying to catch the duck, though why the harriers bothered when the whole plateau swarmed with field rats I cannot imagine.

We saw more Nyala, first four cows and then two fine young adult bulls together. These did not see us and shortly they lay down to chew the cud. They were facing in opposite directions – as Nyala will when they lie down to rest, to avoid being surprised. The nearer one lay in watery sunlight on the side of a small hummock. I had already walked about sixteen kilometres (ten miles) and was tired at that altitude. However, I essayed the stalk.

I had to crawl on my belly for some distance, over bogs, stones, and rocks, cold and out of breath. About twenty-five metres from the Nyala there was a bush of everlastings, the only cover behind which I could approach without being seen. I crept up to it and, raising my head cautiously, saw the motionless tips of his horns. He was looking straight at where I was, and I could not tell whether he was somnolent, or alert, tensed ready to bolt. So I set my camera, focused on the tip of his horns, rose over my bush, and took one breathless photograph before he was up with a snort of astonishment and bolting away across the plateau. He stopped a hundred metres or so away, and let me take two more before he and his companion finally vanished. If I had been lucky I had one epic photograph of a bull Nyala in the open at close range. Needless to say, fear that I had done something wrong tormented me until I could have the film developed. It is not really an epic photograph but it is a good one, and none I ever took gives me greater pleasure to this day.

The two bulls had run off into a deep gully full of long heath, and I thought they might stop there, and perhaps give me another chance. I crept to the edge, and saw them again, with another magnificent adult bull and three cows. The big bull was standing on the far slope near some rocks and was the finest we had yet seen, with splendid lyrate horns. However, heavy cloud had now obscured the sun and photography was useless.

We saw seven more cows on the way back to camp, with a young bull and a calf. In all, we saw 39 Nyala that day. Next day we saw another 18 to 19; and we reckoned that in about fifteen and a half square kilometres (six square miles) there might be 150 Nyala, or one per 10 hectares (25 acres). Evidently, such a density, multiplied over several hundred square kilometres of mountains, could mean a very healthy surviving population, locally abundant. Anyway, it seemed almost certain that the Nyala, believed to number only a few hundreds a few years before, could not be in immediate danger of extinction, and was actually flourishing, at least in Bale. It was almost like what I had read of in Maydon's wanderings forty years earlier, and we rejoiced round the camp fire that night. Mick summed it up: 'We've still got time,' he said; and so it seemed, anyway.

* * *

Mick had to return to Addis Ababa by the next regular flight, and the topography of the mountains, with deep tributary gorges running into the Ganale Doria, made it necessary to drop back down to Goba. In Ethiopia, if you want to do any field-work, the rule is to keep away from officials as long as you can. They may not murder or castrate you, as wild tribesmen may, but they are infinitely more tiresome, with their interminable bureaucratic procrastination, pettifogging obstructionism, and arrogant posturing, than any bloodthirsty man with a knife. Once I was back in Goba, the trouble began. We had gone far too far into dangerous areas, they said. I would have to wait until new mules had been found and an escort could go with me. And so on, and on, and on, through a tiresome morning of argument. Then I had to have lunch with the Governor, out of courtesy, and could not escape until nearly three. I had only got a few kilometres out of Goba and had to be content with a camp in farmland at about 2,750 metres (9,000 feet). It was a pleasant enough spot, and on the right side of the next main gorge; but it was not what I wanted.

I moved up higher next morning and at once ran into more trouble. I was supposed to have been handed on by one local sub-chief to another. However, this gentleman was not to be found, and his people held us up, more or less at gunpoint, while they argued about what to do with us. They stood round in a circle, yelling and shouting at one another, as Ethiopians apparently must, while I resignedly set up my tent in the wet and prepared to wait. Eventually they sent someone off to find the chief, but placed a guard on me and my safari. We could not move, but as it was pouring with rain and visibility was nil because of the mist I was able to write up my notes.

The gentleman who was posted, with his rifle, to guard us was quite a nice old boy. He could not himself see why we could not go out to look for Nyala, but had been told to stop us. Evidently, he would much rather have gone home. We chatted him up, and eventually, when the drizzle eased in the afternoon, he let us go on condition that we returned to camp before nightfall. We did not achieve much in the two or three hours we had, and saw no Nyala. We did see a troop of Olive Baboons at over 3,600 metres (12,000 feet), found tracks of Leopards, and briefly saw a species of Red Duiker which apparently was unknown in this area. The country was very much broken up with steep rocky knobs and deep ravines, and somehow I felt it was not very suitable for Nyala. They were said to come there in the main rains, when the Galla herdsmen moved to lower parts of the forest.

Officialdom relented a little next morning, and we set off to the next main centre of it, at a place called Dinshu, where there was a more important chief. I was told it was a walk of two and a half hours, but should have been warned by the habitual lies and deceit of these people. It took seven hours. We left camp at midday, and at two in the afternoon the soaking rain usual in the Bale Mountains came on. We struggled on morosely for hours until, despite a good waterproof, I had reached a stage of soaking-wet misery in which I would willingly have struck my best friend dead at sight.

Eventually, towards dusk, we came to Dinshu, the usual filthy cluster of tin-roofed mud hovels, with Ethiopians shrouded in *shammas* (cloaks) roosting under the dripping eaves like sodden crows. My men clamoured to be allowed to stay there the night, but the smell and the mud were enough for me. A few kilometres further on I could see a beautiful wood of cedars and, paying no attention to the yells and shouts of protest, I waded through a deep, icy river called the Danka and went on along a rough track. I could not get any wetter than I was, and I was cursing steadily under my breath, anathematizing everything and everybody.

I had not gone far before I saw a short, rubicund young European running after me. He was the Belgian manager of a sheep-ranching concession, François Heck; and he begged me to come and stay with him in his house. He had a good house, he said, with hot water and everything! And the cook from the French Embassy! The thought of a hot bath and a tasty meal I had not cooked myself over soaking sticks made me weaken. With shouts of joy the caravan retraced its steps to the town of Dinshu. My host's house turned out to be a mud hovel only fractionally less filthy than the rest, and lacking a bath; but it did have an iron stove and a chimney so it was at least not full of smoke. François's young wife Christine, far advanced with her first pregnancy, and with a septic foot, welcomed me enthusiastically. So did the fleas, which marched in serried ranks towards the new blood across the floor, undeterred by insecticide sprays. Ethiopian fleas and bedbugs are special! But there was wine, brandy, a good fire, and the cook was as good as was promised; so I was content.

Septic feet do not heal themselves at 3,200 metres (10,500 feet), and I was treating Christine's from my medical kit when the Chief of Police entered and began questioning me. Of course, he had never heard anything about me and would have to send to Goba to get permission before I could proceed. My heart sank, for I knew only too well what that could mean, and I still had most of my expedition in front of me. I could waste weeks there, I knew. However, fate was on my side. As he was writing in his notebook, there was a rifle shot and a chorus of yells outside. Drawing his revolver, the policeman rushed out into the night; and by God's mercy I never saw him again. It transpired later that, in a drunken brawl, a policeman had shot another through the thigh and had dashed off into the mountains, with his priceless rifle, to join the bandits. The entire able-bodied police force dashed off in pursuit. It was the rifle that mattered, not the injured policeman.

My doctoring efforts had not gone unobserved, and shortly after a man came to ask if I could come and treat the wounded man. I had never previously had to treat a bullet wound, and François warned me very explicitly. 'For God's sake don't touch him,' he said. 'If he dies they will accuse you of murdering him. Give any medicines to his wife – then she can be held responsible!'

He went on to recount a similar incident, in which a man had shot his wife through the chest. He had come for two pieces of sticking-plaster, one for the entry hole and one for the exit hole. François had gone along to help, but had given the man himself the bits of plaster to dress the dying woman's wounds. I asked whether she died in the end. 'But of course!' he said with a Gallic shrug. 'What else. But I did not do it, so I was not to blame.' Anywhere else but in Ethiopia one would not believe this story; but there one can safely believe anything.

Humanity, however, dictated that I should have a try. So I went along, to find about forty people crammed into a suffocatingly hot room, with the policeman groaning on his bed. The bullet had passed through the middle of his thigh and smashed the bone. I could do nothing for him but give him pain-killers, and advise that he should be taken to hospital. Warily, I gave the pills to his wife. He did not die; but six months later he was still there, permanently crippled. No one would take the responsibility of transporting him by lorry to Adaba, where there was a mission hospital where he could be treated. If he had died, the relatives would have accused the transporter of murder. Anywhere else I know of, a feeling of humanity would have made someone do something. As it was he was maimed for life.

Next day dawned wet, and continued with unremitting rain. Horses could not have moved in the mud, and I spent most of the day crouching near the fire. In the evening it cleared up a bit, and I went with François to see his new house, a huge stone building being erected on the hillside above by an Italian mason. Nothing was being done by the builder, who was a nice enough man, above 2 metres (7 feet) tall and broad in proportion, with black beetling brows. I should not have cared to meet him in a Naples alley on a dark night. When asked why he was not working he simply said, 'si non piove domani' – if it does not rain tomorrow. He was consuming a substantial meal of spaghetti with unexampled relish. Shovelling it in with both hands and equipped with a set of large, jagged, and powerful teeth in both jaws, he resembled nothing so much as an old fashioned reaper cutting stalks of wheat. A mincing noise, and a fine tomato-coloured spray that hung about his jowls, accompanied his feast. Never did I see anything like it.

François's huge house was planned on a grand scale. It was to have seven guest bedrooms and at least three bathrooms. 'You'll come and stay, I hope,' he said; and I thought I probably would. I asked where the water supply came from, and was told there was a beautiful spring. We went to look at it. A thin trickle came out of a small pipe in a rock-face. It filled an 18-litre (4-gallon) bucket in twenty-three minutes. The Italian builder had no idea of plumbing – why should he after all, he never took a bath himself! François had never thought to test it. That imposing building is still there; but it never was equipped with even one bath.

Next day was better, but still too wet for the mules to move. Mesfin and I spent the day out, looking for a route up the Ueb river valley suited to horses. Of course, I was told there was no way; but we saw horsemen moving along trails and guessed that if we just went we should find a way. That evening François's father-in-law, the concessionaire, arrived with a young sheep farmer from Kenya and his doctor wife. After independence in Kenya he was looking for another home and had been attracted by an advertisement to come and look at this God-forsaken spot. He was almost taken in. 'It's lovely in the sunshine,' they told him. But as the fleas advanced one could see the horror writ large over the wife's visage. I knew who he was, and he had heard of me, but we had never met. Privately I would have seconded the wife's views but felt it unfair to my hosts to say so openly. The father-in-law regarded me askance; he felt I was a bad influence, though he could not quite put his finger on the reason. I said nothing.

Next morning, after a dry night, the sun burst forth in all its glory, and the whole Dinshu Valley was transformed from soaking-wet misery into what it can be – one of the loveliest places on earth. All this time I had been dreading the return of the Chief of Police, his inevitable endless questioning, and the certainty of obstructionism and delay. He had not appeared, and was presumably still wandering dementedly in the mountains in search of the stolen rifle. I guessed it was now or never; and Mesfin agreed. 'Let's go!' he said.

So, despite grumbles from our men, we loaded our four horses and went. As we left the village someone came out after us, yelling, but we ignored him and strode on. We soon left the cultivation and huts behind and came out into the glorious broad upper valley of the Ueb (= wet) River, leading on as far as the eye could see into the heart of the mountains.

All that day we pressed on, never stopping. My only idea was to put as great a distance between me and officialdom as possible. Mostly we travelled on a good track, where we made good speed, but sometimes we had to wade through bogs that gripped us to the waist. On another day I should have liked to stop and look at those bogs more carefully for they were full of the endemic Blue-winged Geese and Rougeot's Rails, and I also saw several pairs of Ruddy Shelduck, and suspected that they might even breed there. However, the fear of officialdom on our tail was a spur to unremitting effort, and we never stopped until we had put a good distance between us and Dinshu. Looking back along the trail I could see no signs of pursuit for a long way in the sunlit evening. We camped on a dry hillock, with plenty of dead heather for a fire, feeling that we had seized our chance, and won.

A few days later we were not so sure. We moved camp across the valley to the foot of a big cliff, and here, after our record and non-stop trek, we rested all one morning. In the afternoon Mesfin and I climbed up above the cliff and through a most extraordinary region of contorted lava flows. The mist came down and swirled about us. It was cold and damp, and though we found tracks we saw no Nyala. This inhospitable terrain swarmed with Rock Hyrax and there were plenty of hares, but we were in danger of losing our way entirely in the fog and being benighted up there. So we turned back for camp and, walking very quietly in case we came on Nyala, reached the edge of the cliff. There, looking down on our camp, were three shrouded human figures. With their backs to us they looked full of menace. We had little doubt that they were some of the dreaded bandits of whom we had been told so much, who were infesting the mountains, and who were a plausible reason for the difficulties put in our way by officialdom.

They had not heard our approach. I looked at them through my binoculars and, as far as I could see, none had a rifle, which a seated Ethiopian usually holds with muzzle pointing upwards over one shoulder. We did have a rifle – taken so that we could shoot a duiker for food if we could not buy a sheep – but had left it in camp. Mesfin, thinking of his manhood, looked a little grey about the gills. We were separated from them by about forty metres of nearly bare rock slabs, so that in rubber-soled shoes we were able to approach in dead silence. If any of them had turned things might have been different; but they were completely intent on our camp directly below them. They had no guns and had left their spears leaning against a heather bush a short way behind them. It was simple to move forward the last few paces, and grab their spears. They were startled almost out of their wits but, with a vertical precipice in front of them, could go nowhere.

We explained that we were only scientists looking for Mountain Nyala; and once the first anxious moments had passed it all became quite friendly. They came to camp, had tea, and agreed to provide local guides for us the next morning, but said that there were few Nyala in this rocky terrain. We parted cordially; but I did not sleep very easy that night, and deliberately moved my tent after dark to another spot further from the camp fire. Nothing happened, and in the morning a guide duly appeared. Later in the day we met a local sub-chief, one Kadir Haji Biftu, a very pleasant man in his middle thirties. He said, yes, there were many bandits in the valleys to the south, but none here. The people we had taken for bandits were local herdsmen who, alarmed by this sudden invasion of what they thought might be Government troops, had come quietly to look us over.

We were now blessed, after the days of soaking rain, with a week of glorious weather and we went far afield each day. One day we went up a valley between two lava flows, each ending in impressive columnar cliffs. At first it was broad, then it narrowed, and finally became a defile less than fifty metres wide overlooked by vertical cliffs on either side. As I walked up that defile I never in my life felt more like a sitting duck. Emerging into a slightly more open place, where the encircling cliffs drew back a hundred metres on either side, I saw a head peeking over, and my heart almost stopped. Binoculars have their uses, however, and, on looking closer, I soon assured myself that it was nothing worse than an Olive Baboon.

We went on and on through the lava flows and found one big Nyala bull feeding in a patch of burned heath. The light was poor, but I managed to stalk and photograph him. When I was within fifty metres of him a puff of the uncertain wind gave me away. Instead of bolting, he rose and withdrew quietly into the dead burned stems of the heath, where with his dull grey coat, white stripes, and facial chevrons he was perfectly camouflaged. Here he allowed me to take several photographs before finally deciding that he should leave the area. I last saw him going over the opposite ridge at a full gallop, bounding, outlined against the sky.

We dropped into another valley, where I found a Verreaux's Eagle's nest, at almost 4,120 metres (13,500 feet), with a nearly mature young bird in it. It was the first definite breeding record for Ethiopia, and surely an altitude record for a nest of this species. Although there were plenty of hyrax, their favourite prey, here about they evidently also killed Klip-springer, for we found the remains of one eaten by them. It had been an adult; and its horns still hang in my son's bedroom. A little further on we emerged on to open grassy moorland, much more like country for Nyala, but we saw none, though we did see a pair of Blue-winged Geese at 4,000 metres (13,100 feet) by my altimeter, and three Semien Foxes.

At this point the upper edge of the lava flows through which we had been climbing all day had been weathered into the most astonishing rock formation I ever saw. The molten rock had cooled in columnar formation, as volcanic outpourings often do. These columns, each about two metres (six feet) thick, had been weathered into rounded blocks, with narrow passages between, full of hares and Klipspringer, safe where no Leopard could reach them. The winds and the rain had weathered one side of each upright block more than the other, so that they looked for all the world like those ice formations in the Antarctic, created by wind, called 'penitentes'. In fancy they resembled a vast crowd of brown-clad monks, cowled, and standing with their backs to me. I tried working my way in among them, but could not get very far in because the cracks between were too narrow, and I could not jump from top to top because the blocks were too far apart and of uneven height: it would have been a bad place to break a leg. I felt pretty sure that no other European foot had trodden the tops of those astonishing columns. It is one of several places that I would pick, for a student I especially disliked, to make an ecological survey.

We walked a good 40 kilometres (25 miles) that day over some of the roughest and wildest ground I ever trod. We were actually at the head of a great gorge, one of the headwaters of the Ganale Doria, into which Bill Corcutt and I had looked daily in 1963. We had then christened it the Ganale Chasm; and chasm it was and no mistake, falling away for some 900 metres (3,000 feet) of rolling ridges covered with extensive Hagenia forests, and with little green alps nestling under beetling cliffs along the ledges of its slides. It looked an intriguing bit of country, really wild and remote. However, I was never to get into it because even Kadir Haji Biftu said it was full of *shifta*, and all of my men refused to go there at all.

From here Kadir Haji Biftu led us to a camp at over 3,600 metres (12,000 feet), where it was piercing cold, but where there was a dense population of Nyala. In two days we saw 43. It was clear that Nyala were locally common but that one might go for days without seeing one, if one did not know where to look. Here John Blower, wildlife adviser to the Ethiopian Government, joined me. And then we moved on back to my old 1963 camp above the valley, where we had lived in the Galla huts and been surrounded by the spearmen. Here we saw more Nyala, including one magnificent bull, possibly the finest I ever saw, maybe even the same one who had given me that unforgettable moment in 1963, though that seems unlikely. He came nosing out of deep heath with extreme caution in the evening and stood for many minutes stock still before finally satisfying himself that all was completely well and that he could begin to feed.

Here also we lost Mesfin. We had sent him off to do a survey of an area on his own, as part of his training, and he did not return. We searched for him all of one day, but could find no trace of him. So, for the time being, we had to abandon the expedition and descend to Adaba to report to the authorities. It was a 27-kilometre (17-mile) walk, mostly downhill along good tracks; and I shall never forget the manner in which John Blower walked ahead, not deigning to speak to his companions, at the pace of a forced march. I cannot think why it was either necessary or desirable. In the end we found Mesfin in a hotel at Adaba. He said he had been attacked by bandits, robbed of his money, and had to flee for his life. Whether that was actually true, or whether he had finally got fed up with the cold and the wet and the hard slogging work of continuous field surveys, I never found out. However, I took him at his word. He rejoined the expedition and we finished the trip together.

I had now been living at 3,400 metres (11,000 feet) or over for a month. I had seldom walked less than 24 kilometres (15 miles) in a day, and on the heaviest days had done well over 32 kilometres (20 miles), sometimes reaching over 3,900 metres (13,000 feet). The horses were done and I had to hire fresh ones. We all needed a break. So I went down to Lake Langanno and met Emil Urban for another trip to Pelican Island, leaving Hussein at Dodola to rest, obtain new horses, and be ready for me when I rejoined him four days later.

I was already sure enough of my results. Up to that time we had seen 131 Mountain Nyala, of which 25 (19 per cent) were males, 78 (59 per cent) females, and 19 (15 per cent) calves; the remaining 9 could not be certainly identified as to sex or age. Of the 25 bulls seen 11 were mature animals with 'shootable' heads; and of these I had actually photographed three. About one female in four had a calf at foot. Thus it was abundantly plain that the population was healthy, locally common, with a good scatter of adult bulls and a satisfactory reproduction rate. The Nyala, whatever else we might later find, was certainly numbered in Bale by the thousands, and not by a few hundreds, as had been feared. I still had work to do, but had broken the back of it.

I returned to Dodola on 1 March. I reached the town at about midday, and there was then a long wait while Hussein collected the horses he had managed to find. Believing that it was best to get going while we could, we made a short trek and camped 8 kilometres (5 miles) from Dodola that night in cedar forest. Next day we crossed the high ridge behind

Dodola, passing west of the Lammergeier's nest I had found in 1963, and made a nice camp at the upper edge of the forest on the southern side of the mountains. The horses bolted *en route* and my boxes and stores were scattered all over a grassy glade, but we caught them, put all the baggage together again, and pressed on to camp. As usual, in the afternoon it was misty and drizzling, but I saw a herd of Nyala cows near camp, and felt that I was well placed to find out what was to be learned of the remainder of the main Bale massif.

We were on the southern side of a big mountain called Corduro, and not very far from a military encampment. The ex-District Governor of Dodola, a formidable old man whom I would not have liked to cross, had actually been killed by the *shifta*, though in 1963 he had told Bill Corcutt and me, with a grin, that there were no bandits in *his* mountains. As a result, military activity was considerable, and aircraft passed over daily. Mesfin and I separated each day, I going with the head horseman, named Tuke. I could speak only about ten words of his language, and he none of mine, but we agreed well. He was a natural hunter, and very keen-eyed. Somehow we could understand each other by a few signs without having to speak. I have met many others like him in my time. Mesfin went with Hussein and had now reached the stage when he could do a day's survey by himself.

Two memories stand out especially from those few days, both concerning individual bull Nyala. The first was when Tuke and I, sitting on top of a rocky peak in full sunlight, had thoroughly and minutely spied the whole of a big valley beneath us. Had there been any Nyala moving in it, we must have seen them, but there was apparently nothing. We had been sitting in the open on the top of the rock for twenty minutes or more, occasionally making a remark in our very limited conversation. Then, when he thought we were not looking his way, a huge Nyala bull rose from a small patch of heath in the middle of a little open glade just beneath us, and sneaked quietly away, with his horns laid back on his withers, until he was in the shelter of the dense heath. If I had not happened to glance in his direction at that moment I should never have seen him move. How many more Nyala were there in that silent empty valley? Lots, would be my guess.

The second bull I photographed. We saw him emerge into a small patch of burned heath, in full view, but managed to crawl into cover before he could see us. After grazing for a while, he lay down. To give you a clear idea of his native cunning, he had at his back a patch of dense heath through which no enemy could approach him unheard. He lay looking at the patch of open ground where he had grazed. And on his left, where he could have been closely approached beneath a small cliff, he was protected by the wind blowing gently up valley. If he had reasoned it out, he could not have placed himself more cleverly.

Without any difficulty I stalked him to within thirty metres, from where I could see the tips of his horns, moving gently as he chewed the cud. I could not photograph him because of the intervening dead stems of giant heath. So I rose slowly into his view and, just for the record, photographed his surprised visage as he too stood up. Then he gave a 'Hrrmph' of alarm and bolted into the valley. I expected him to continue without a stop, but he was not sure what had startled him, and he stopped a hundred metres away in quite short green heath and let me take two more pictures. He was a fair trophy bull; and, if I had had a rifle, he could have been as dead as mutton, had I wished. I hope he learned a lesson, for no Ethiopian poacher would have been as forbearing.

When Mesfin and I left the Corduro massif we had covered all but an outlying spur of the Bale Mountains and some parts that were inaccessible because of the bandits. We had seen between us 196 Nyala, and had made good estimates of the age and sex structure of the population. We had formulated plans for a National Park in outline, and were certain that it would contain at least 2,000 Nyala, besides a thriving population of the Semien Fox, any number of endemic birds, and lovely scenery. The only thing against it was the climate; but we had had a fair share of good weather too, and I reckoned that horse-trekking, from one well-found hut to another, might prove popular. I had great hopes.

There remained the Arussi Mountains, the place where the Nyala was first discovered, and the great hunting-ground of the twenties and thirties. I knew that most of the forests on the lower slopes, virgin in Maydon's time, had since been destroyed by cultivation. However, there seemed a good chance that there would be a sizeable population of Nyala on the higher heath slopes. To reach them, we had to hire new horses at Dodola, for Tuke and his team would not come; and here we again ran into official obstructionism. The local Governor, a sneaky little man called Balambaras Aseffa Wondim, refused to help unless we paid an extortionate price in advance. So we finally induced Tuke to take us to a little village called Asasa, in the middle of the open Uebi Shebeli plain, where there was a delightful warm spring swarming with birds. Emperor Haile Selassie used to come to a private airstrip there, and a thatched pavilion had been built for him from which to watch the birds – which he enjoyed. I enjoyed it too, while Mesfin negotiated the hire of four more horses.

We were by now all fairly tired. However, I felt that it was absolutely necessary to complete the survey as best I could. So we climbed the highest of all the mountains, Mt. Cacca, and then went over the Galama ridge, at the southern end of the main Arussi range. This time I rode a horse; but I was then a big and heavy man, and Ethiopian horses are small, if wiry. At 2 p.m. I rode into a small village, where it was market day, with hundreds of people about. I bowed right and left in gracious acknowledgement of the courteous salutations of these people; and then my wretched horse decided he had had enough. He simply lay down under me, and I had to dismount in a very undignified way to avoid being rolled upon. Ethiopians are all good horsemen, and it simply made the day for them. They were hysterical with laughter, and I could see it was funny too.

We camped that night near a sacred grove of cedar trees, and as a result could get no firewood, for the locals would not allow us even to pick up dry sticks. Thereafter we traversed along the top of the Galama ridge, keeping high up as we had done before. We found very few Nyala up on the ridge, but we saw one herd of thirteen accompanied by a magnificent bull, who ran off by himself when startled. There were too many people about, and half-way along we had to descend to the upper edge of the cultivation for provisions. It was a very long walk indeed and I was dead beat that night. When it poured with rain the next day I was quite happy to stay in bed, scarcely moving, all day long, sleeping most of the time, but eating an odd biscuit and drinking tea laced with whisky. I had had about enough of this expedition by now and decided that one more high camp would have to do.

Next day was fine, and we climbed up past an impressive volcanic plug called Boralugu, which I could recognize from Maydon's old descriptions as 'Robber's Mountain'. Here he had found plenty of Nyala, but we saw only two young cows all day. I wanted to make camp on the south side of the range and, since no one else knew the way any better than I, rather misguidedly crossed a deep valley instead of going right round the head of it. We reached our chosen camp site after a fearful struggle, for there was a little cliff hidden in the giant heath, only about three metres (ten feet) high, but a near-impassable obstacle for laden horses. Camp was at 3,800 metres (12,500 feet), and extremely cold; but there was some grazing for the horses, and it would do.

Up to now it had been my practice to go out with one horseman or with Mesfin. On this trip we had with us a horseman called Abdurrahman, who seemed more intelligent than the others, and who was therefore called on to come out more often, while others stayed in camp. On 17 March, our second day at this rather uncomfortable camp, he and Mesfin and I went up the ridge towards the highest peak in this part of the range, Mt. Badda, about 4,150 metres (13,600 feet). At 4,060 metres

(13,300 feet), on the top of a cold windy ridge, Abdurrahman hurled himself to the ground, apparently screaming in agony, with real tears starting from his eyes. 'God!' I thought. 'He's going to die on us!' We could do little, but laid him down in the shelter of a bush, while we debated what to do. As we were discussing whether to try to carry him back to camp I noticed that, between groans, he was looking at me surreptitiously out of the corner of his eye. I had not been in Africa for thirty years for nothing; and it struck me suddenly that he might be shamming.

I could not at first think what to do. Then I had a brainwave, and took his pulse, reckoning that anyone who was suddenly struck by a painful seizure would have a fast heartbeat. It was absolutely regular, bonk-bonk-bonk, little above the standard 74 a minute, even at over 3,900 metres (13,000 feet). Shamming he was, but I could not convince Mesfin of this, and the last I saw of them was the couple staggering back to camp along the ridge, Abdurrahman draping himself about Mesfin, who practically carried him all the way.

I went on and did the rest of that day's survey alone. In those wild great mountains, where an armed Ethiopian seeing a lone European could very well have pulled trigger first and asked questions afterwards, I felt more lonely than ever before in my life. I kept a sharp look-out not only for Nyala, but for any human being. Long years of experience in avoiding the unwanted attentions of Scottish gamekeepers stood me in good stead. I climbed to the top of a mountain where my altimeter read 4,102 metres (13,460 feet); it was almost certainly higher. Apart from the height, I could have been in a wild part of the West Highlands, from the shape of the hills and the generally open, sparsely vegetated tops. However, the Augur Buzzards, the Lammergeiers, the giant lobelias, and far away and far below the gorges at the headwaters of the Uebi Shebeli running towards Somalia told me where I really was. All day I saw no living soul, nor any Nyala, though I found a few tracks, and saw odd Klipspringer and duiker.

In the afternoon, near camp, I rested for an hour in baking sunshine near a small swamp. As I lay there a Semien Fox, the third I had seen in Arussi, came out of its lair and proceeded to hunt rats in the rank vegetation. It found a nest of something, and ate the contents, then went on, making an occasional pounce with its forefeet, and wagging its tail until it lay down again two hundred metres further on. I rose and made my way back to camp, cheered by the sight of this pretty animal going about its business as if I were not there.

Abdurrahman was still alive when I got back to camp. When asked if he was any better he said 'Tinnish' (a little), in a quavering voice. His pulse was steady and strong, and he had no temperature, but he complained that he could eat nothing, and had a pain both in his belly and in his head. Mesfin was still convinced he was sick, but I had my doubts.

Next day I took it easy on a sunny morning, and had my photograph taken for the *National Geographic* bathing naked in an icy stream. In the afternoon we went out separately, leaving Abdurrahman in camp. He had still been publicly unable to touch a morsel of food, but while we were out he raided the other men's stores and stuffed himself. When we returned he was moaning by the fire and covering his head with his cloak. The other men were outraged. 'There's nothing wrong with him,' they said. 'Make him do his work.'

On 19 March it was warmer, and a change in the weather looked imminent. I had no desire to be caught up there in torrents of rain, and I reckoned that we had made a fair survey of most of the Arussi ridge too. So we packed up and went. I asked Abdurrahman whether he was fit to work, but he said he was not. So I told him to ride my horse, and he did so, swaying from side to side like a dead body, head hanging, all the way over the very top of Mt. Badda and down the other side for 27 kilometres (17 miles) to a valley at 3,000 metres (10,000 feet), where we camped in the evening. At that altitude the air seems rich and warm after 3,800 metres (12,500 feet); and we sat round the camp fire and talked. I thought the time had come to spike Abdurrahman.

'Tomorrow we shall get to Asella,' I said. 'And since you've been so ill I shall take you to the Mission Hospital and ask them to cut you open, to see what's wrong with you.' Amid a hush, Abdurrahman said that, as a matter of fact, he now felt so much better that he did not see the need for that. 'In that case,' I said, 'you can either admit you've been shamming these last few days, or I shall certainly take you to the hospital and insist they cut you open.'

Abdurrahman decided that his time had come. Yes, he had been shamming, he admitted; he had not felt like coming out with me all day. 'Right,' I said. 'In that case you will not only be fined for the days you have not worked, but I shall charge you for the hire of the horse you rode today.' It went right home in his vitals. There was a concerted roar of laughter from the other men at his discomfiture. Justice was not only done, but was seen to be done, and no one took any harm from it. I solemnly paid Abdurrahman his wages and then deducted one dollar and fifty cents for the hire of the horse.

We slogged into Asella the next day. We had walked an estimated 1,030 kilometres (640 miles) in forty-eight days in the field, right along the tops of the Bale and Arussi Mountains, most of it in rough country at over 3,350 metres (11,000 feet). I was tired as I never have been tired in my life; bone weary all through. However, I had done what I came to do, made a reasonably thorough status survey of the beast I sought to study, and concluded that it was not in any danger at all in the Bale Mountains, which would be the right place for a National Park. Even in the Arussi Mountains we had seen 35 Nyala, including 4 bulls, 20 cows, and 11 calves – a higher percentage of young animals than in Bale. Given any sort of protection from continuous hunting the Nyala would survive for ever, if human beings could be prevented from destroying its habitat. It was the longest and hardest walk of my life; but it was well worth it.

Since then, a National Park has been planned for most of the Bale Mountains. On one easily accessible mountain called Gaysay, opposite the big house at Dinshu, which I did not climb in 1966, I myself saw 41 Nyala in a day on 20 November 1971, beating the record I set up with Mick Prosser in 1966 above Goba. They included the only pair of twin calves ever seen in the Nyala, and one big bull. Since then up to 70 Nyala have been seen in a day on Gaysay Mountain, in an area of only about eight square kilometres (three square miles). If protected, they can evidently become very numerous. Gaysay is ideal habitat, with open grassy areas alternating with patches of Hagenia and heath for cover; elsewhere densities would not be so high. However, I calculated in 1966 that at worst there might be 4,600 Nyala in the Bale Mountains alone, and at best over 11,000. The true figure may be somewhere between the two. However, the full range of the Nyala to the south, in Sidamo, is still unknown, and in those Hagenia forests which I saw in 1966 and could not visit because of the bandits there may be many more.

The big house at Dinshu was abandoned by François and his wife when their sheep-ranching enterprise failed. It became the headquarters of the proposed Bale Mountains National Park inhabited by several Peace Corps biologists. In 1967 the Danka and the Ueb rivers were stocked with Rainbow and Brown Trout respectively, and this has resulted in superb trout fishing, some of the finest in the world. I have only fished there three times but never failed to make a good basket, in the most glorious surroundings.

The last time I was there was on the week-end of 22–4 November 1974. I fished for one evening and, using the same fly throughout, hooked about twenty-five and landed sixteen Rainbow Trout, weighing 340 to 570 grammes ($\frac{3}{4}$ to $1\frac{1}{4}$ pounds). Next day I caught four beautiful big Browns in the wonderful clear pools of the Ueb, on a small dry fly of my own invention. My wife went to Gaysay Mountain and there saw some Mountain Nyala, all on her own. We collected a red *Acanthus* that now grows in our garden at Karen. Mercifully we had taken no wireless, and so were unaware of the bloody events taking place in Addis Ababa. It was the night when the extremists among the ruling Military Council murdered the Head of State, the moderate General Andom. For good measure they killed fifty-nine other political prisoners, even wheeling out old men dying of terminal cancer, and machine-gunning them against a wall. On our way back to Addis we were stopped and searched; and from then on things seem to have generally gone from bad to worse in Ethiopia.

The Bale Mountains National Park has never been legally established. The area was again in dispute in 1977–8, between Somalia and Ethiopia, as it was in 1966; and no doubt the mountains were full of armed gangs again. However, the last I heard, in 1977, was that the Nyala were still flourishing on Gaysay and elsewhere. Perhaps I can take some comfort from the head of the Wildlife Conservation Department in Addis Ababa, whom I saw briefly on my way to detention in Somalia. 'In these days,' he said, 'people have no bullets to spare for wildlife, only for each other!' Well, it is a grim thought; but maybe the Nyala will survive it all, and some day the Bale Mountains will become a tourist resort. I shall probably never see them again; but I saw them once, all the way along the top of the mountains.

Standard-wing Nightjar
Pennant-winged Nightjar
Potoo
'poor-me-one'
D. TISCHLER
Cayenne Nightjar on nest

9. Nightjars at Night

Many people are afraid of the night. This is quite natural for, descended as we are from diurnal primates that roosted at night in trees or crouched in caves for fear of nocturnal carnivores, we have excellent binocular vision by day, but very poor night vision. With enough practice one can learn to walk silently through even quite a dark wood by night; but one does it by feel and touch almost as much as by sight. On open moorlands or salt-marshes one should have no difficulty on a moonlit night, but on a dark cloudy night with no moon it is not so easy. I never believe anyone who says he can find his way about by night as well as by day, because I know I cannot, and I have done a lot of it in my time.

The best cure for fear is to face it. If frightened of the dark, go out in it, and get used to it. To the inexperienced ear, the sudden outcry of a creature one does not know makes the heart go pit-a-pat. Abroad at night one hears noises that are at first inexplicable, and can be alarming, much more so in Africa than in Britain. What is that stealthy footstep – an antelope or a Lion? If you can hear it, probably the former, for a Lion that wants to eat you will not advertise its presence. What is that stealthy rustling, a venomous snake, or just a harmless mouse, even a beetle? The answer is, find out.

Take a powerful focusing torch and, when you hear any inexplicable sound, pinpoint it as best you can by turning your head this way and that, and then illuminate it. You can then see it, and will almost automatically lose any fear of it, though a gentle rustling can turn out to be a bull elephant. Nocturnal animals usually do not expect to be suddenly flooded with light, and are confused by it, so that they do not at once flee. A hare or rabbit caught in the headlights of a car is a case in point; but you will not find a domestic cat, a true creature of the night, confused in the same way.

When I was young in Scotland I spent much time night-walking, especially on long clear cold winter nights, occasionally under the flickering glow of aurora borealis, which we call the Merry Dancers. I came to enjoy it by being often about at dusk, when the creatures of the day go to bed and those of the night emerge. Briefly, both may then be seen or heard. Waiting quietly in a wood, I have seen all sorts of creatures, unsuspecting. I once had a Fox trot right up to me, his footsteps clearly audible on leaves crackling with frost. However, unlike those Badgers in Oxfordshire, he at once grasped what that still figure leaning against a tree might be, and trotted no more. Badgers and Foxes are both mainly nocturnal; but one has poor, the other good, night vision.

In the tropics, wherever one is, night-walking is undoubtedly more dangerous than in Britain. Most venomous snakes are largely nocturnal, and are also then more active and actually dangerous. Snake lovers are inclined to say that these dangers are exaggerated, but the fact is that most people who die nowadays of snakebite are barefoot people who walk about at night. Having been brought up with a healthy respect for snakes in India, when I move about at night in Africa I take a torch, and tread carefully and warily. I examine any long thin object across my path with care. All the same, there is a magic about the tropical night which is undeniable. It is often still, warm, scented; and if the moon shines it shines so brightly that one can even read by its light. Such a night entrances many a newcomer, and imbues him or her with a lack of caution soon to be regretted. The man who rushes ecstatically, naked, into the moonlit African night, seeking the embrace of Mother Nature, will more likely find himself in the clutch of a Wait-a-bit Thorn. Nothing is quite so disillusioning as trying to disentangle oneself from the tenacious grasp of those clawed twigs, when one cannot even see which way to pull. Our worshipper returns to camp, bloodstained and chastened, a sadder and a wiser man.

Any night-walker will soon learn that many birds, even many that might be thought diurnal, are abroad and active by night. In northern estuaries herons and waders are on the move and much small bird migration takes place at night. Flamingos move from lake to lake in the Rift Valley flying high up at night, when they are safe from both diurnal eagles and from owls, which do not fly high or soar. Good night vision is not absolutely essential – both ducks and flamingos see less well by night than by day.

Some birds, however, are specially adapted to the night. They include both owls and nightjars, the latter even more purely nocturnal than owls. Some owls move about freely, even hunt, by day; nightjars seldom or never. By day one usually sees them only when they rise from almost beneath one's feet, to flit a little way and settle again, often in the shade of a bush. Nightjars first become noticeable at dusk, and are active through most of the darkest nights – so far as anyone knows. Owls have been quite intensively studied by many people, despite the difficulties; nightjars by very few. Perhaps this is because owls have forward binocular vision and fluffy plumage, which is supposed to make them appealing. They have been thought of as birds of wisdom; or alternatively as witches' familiars and as birds of ill omen, either to be actively persecuted, or just to be avoided at all costs. These superstitions speak of our own fear of the night, and of anything that is unnatural enough to move, hunt, and hoot sepulchrally in it. The nightjar also is the victim of silly superstition. It is – or was – believed by Europeans (never by Africans, who know better) to suck the milk of goats – hence the alternative name 'Goatsucker', translated into the Latin generic *Caprimulgus*. One wonders, why goats only? Something to do with Pan, or Satan? Evil and mysterious anyhow. Yet a nightjar is a harmless, indeed a beneficial bird, totally inoffensive to man.

At night ears, both for humans and other creatures, come into their own. Nocturnal animals often have very keen hearing, and signal to one another largely by loud and distinctive noises. Lions roar, the tiger walks the jungle uttering his bass mew, hyenas howl, owls hoot, vixens scream, not to frighten prey, but to tell others of their kind where they are. The Tree Hyrax of East Africa, silent by day, emits a terrifying volume of sound for so small an animal by night. It can scare the greenhorn rigid; but a torch reveals an inoffensive, rabbit-sized animal sitting on a branch, making a noise like a cross between a steam whistle and a football fan's rattle. Sometimes one cannot locate the unknown sound easily, even with torchlight. It is very hard to see the little African Scops Owl, whose familiar trill is beloved of all those people who camp in the bush.

Like other night creatures, nightjars are vocal. They call at dusk and dawn on dark nights, but in moonlight will go on all night, without cease. One wonders how they draw breath. Once sitting up for a Tiger in India, I heard a Jungle Nightjar calling all night long from a rock outcrop. I slept intermittently, but whenever I woke in the moonlight the bird was still emitting its monotonous 'Chuckoo-chuckoo' without cease. At Karen, if I wake of a moonlit night, I inevitably hear the shrill beaded whistle of the Abyssinian Nightjar; and in the Kenya thornbush the dull 'tuk-tuk-tuk' of the Slender-tailed Nightjar goes on all night, under a moon. No diurnal bird sings as incessantly as a nightjar. I suppose they eat sometime, and do not only feed on dark nights, but they usually only stop singing or calling when they are in flight.

A nightjar seen by day presents a problem. Most will allow very close approach, then rise and flit a few metres to settle again. Through binoculars, on the bare ground or among leaves, they still all look much alike. What seems to be good field marks, such as white bars on the wing in flight, or white tail feathers, turn out common to many species. When you consult a book you are often none the wiser, for the distinctions clear in a museum skin often do not show up in the field. So you often remain ignorant, unless you shoot your nightjar; even then, you cannot be certain that you would know it again if you saw another.

So, to learn about nightjars, you must become a creature of the night yourself, and use the same means as they do to distinguish another – voice. Ears are far more useful than eyes for identifying nightjars; but, unless you want to kill it (which I never do), you must first learn to catch your nightjar. Then you can connect the voice with the bird; and after that it is easy.

I first became interested in nightjars in Trinidad, in 1940. In those days I was basically a bird photographer, but on that tropical island where so much could still be learned so much more easily than in Britain, I soon found myself making odd new observations. Other ornithologists told me to concentrate on vireos, or small forest flycatchers, but when young one tends to study what attracts one most, and I was not interested in small dull birds, though bell-birds and motmots, almost equally unknown, intrigued me.

Oddly, one of the birds that had been well studied was a nightjar, but a queer one. It was an arboreal species, the Potoo, *Nyctibius griseus*, known locally as 'Poor-me-one'. The name derives from the exquisite liquid whistling call, that shames the Nightingale in a French wood and the trilling Curlew on Scottish moors in spring. It means 'Poor me! All alone', and the five pure, soprano notes, ending in a kind of chuckling sigh, fall on the ear so sweetly that one feels one must rise from one's bed and search for the author. By day, the Potoo imitates the tip of a grey, lichen-covered stump, sitting bolt upright, eyes closed to slits, beak pointed skyward. I once heard this lovely call on a moonlit night; and maybe that is what started it.

Nothing seemed to be known of other Trinidad nightjars. The two commonest species, which I sometimes heard at night, made calls like 'Who-are-you' and 'Chuck-will's-widow'. Several American night-hawks, as they are called, seem to make calls of this sort. Sometimes I flushed these birds from the dry leaves of the bamboo brakes near the Imperial College of Tropical Agriculture, but I could not tell which was which without handling them.

Above the Imperial College, on the ridge leading up into the forest beyond the Mount Tabor monastery, there was a delightful patch of open savanna. Here I often went for a walk because, up there, the sun never shone unmitigated without a cool breeze, there was the shade of the deep forest beyond, and when I came down again in the evening there was always a magnificent view of the whole island and the Gulf of Paria beyond. It was a lovely place.

One afternoon, as I was coming down, a friend coming up told me that if I searched near a certain white boulder I would find a nightjar's nest with an egg. He had flushed the bird by chance, at his feet. I soon found the egg, but of the bird there was no sign. She may have been watching me from a patch of bare ground not far away, but I did not search for fear of frightening her more. I laid a few branches beside the nest and departed at once.

In my bird photography I had been using flashlight apparatus, of a very much more primitive kind than is available nowadays, to photograph humming-birds in the deep gloom of the tropical forest and the unique Guacharo or Oil Bird in the limestone caves. I had taken a liking to the method, because of the beautiful sharp results it produced. Here was a chance for me to try working at night, to study a species of bird more than usually still and wooden-looking by day. It also seemed likely that no one else had ever photographed this species of nightjar, and that little in detail would be known about it. I could be breaking new ground.

Looking at my nightjar a few days later as she sat on her eggs I had the usual problem, complicated by the fact that in those days there were no handy reference books. In a museum skin, all the local species could be distinguished by such things as the amount of white on the throat and the extent of a rufous half-collar. When she sat crouched up, her chin drawn in, and her head sunk into her shoulders, my nightjar did not clearly show any of these characters. So when I first went to photograph her at night I still did not know what species I might be photographing. She would allow me to approach to within about a metre and take her photograph by day, but then sat absolutely still, like a mottled stone. If I moved slowly round her I could see the great eye on the far side slowly closing, and that on my side slowly opening, like the waxing and waning of a glowing spark. Other than her eyelids, she never moved a muscle while I built my hide close by and took daylight photographs.

Of the five species in Trinidad, one was the forest-loving Potoo, and I eliminated one other possible species on the grounds of egg-colour and the size of the bird. My bird might be any of three species, the Who-are-you, the Chuck-will's-widow, or the very little-known Cayenne Nightjar. Only four nests of the Cayenne Nightjar had ever been found in Trinidad and almost nothing was known of its habitats. Although I fully expected that my bird would turn out to be one of the common species, I was excited at the possibility that it might turn out to be the Cayenne Nightjar, whose calls were undescribed.

When I settled myself in the hide towards dusk the savanna was sweet with the whistling calls of *Elaenia* flycatchers and of tinamous in the bush around. The nightjar had left when my cumbersome flashlight reflector was attached to the roof of the hide, and in leaving had exposed a single newly hatched chick. The tinamous and the flycatchers grew silent, and lights began to pop out in St. Augustine on the plain below. Then, when it was still too light to use the flashlight effectively, my bird returned.

She alighted a little way beyond her chick, and with wings upheld ran towards it with little tripping steps, uttering soft 'wut-wut-wut' cries. I had never seen her before when she did not realize I was there. I was entranced by the transformation from the block of mottled stone, immobile on the ground, to this graceful creature advancing, in a movement almost ecstatic, towards her chick. She settled before I could take her photograph, but was no longer still like a stone or uninteresting. Alert, but thinking herself unobserved and so at ease, she became fluid and flexible, bobbing her head up and down. With her huge dark liquid eyes wide open, it seemed as if something dead had become alive. She was transformed from just a marvellous example of camouflage into one of the most lovely birds I had ever seen.

I decided to let her sit at ease until it grew fully dark before photographing her. Occasionally I shone my torch on her, and she paid no attention, though her wide open eye threw back the light. Then, for a reason unknown to me, she suddenly left. Quite startlingly loud and near there came a piercing sweet whistle, 'Chi-peeeeeuw', very long drawn out and sad-sounding, but pure and true. With that sound I knew instantly that my third unlikely possibility was true and that it was the Cayenne Nightjar I was watching, not the Chuck-will's-widow or Who-are-you.

Time and again that piercing whistle sounded, apparently from the top of the white boulder. I assumed it was the male who was calling, answered by the female with a low subdued cry. Then both took flight over the savanna together, uttering soft conversational 'wut' calls. A moment later there was a flutter and a bird alighted beyond the chick. I could not focus any better in the dark, but thinking to take a picture of the female settling I pressed the trigger. In the brilliant light of the flash-bulb I saw the male nightjar, markedly whiter than his mate, sitting just beyond the chick. Chagrin at taking a picture certain to be out of focus was mixed with delight at confirmation, by this useful field character, of the identification of my bird.

Later I took several photographs of the female, none quite perfect, but good enough. The chick grew very rapidly, and soon became active. Once it could move the female no longer came to the nest, but would alight beyond the chick and call it to her with soft 'wut' calls. I tried tethering it to a stone with a belt of golf-ball elastic, but that slipped off; and soon I had to give up taking pictures from the hide. The male came only once more to the nest, and I failed to photograph him then. Each night he alighted on the white boulder soon after dark, and uttered his piercing sweet whistle. Sometimes the female joined him, and sometimes he flew away, and I would hear him calling in the moonlight over the savanna.

As this chick was now too active to stay in the nest, I contented myself with taking photographs of it, focusing the camera in the powerful light of a torch. Very often the female sat beside the chick, and I found that I could photograph her too by torchlight. She would even, when dazzled by the torch, let me stroke her back. I could push the torch right up to her, reach out and touch her; and if I had not felt that this might terrify her and make her desert her chick, I could have picked her up. I never got a perfect photograph in this way, however, because whenever I switched out the torch she immediately blinked, and every picture I took showed the nictitating membrane drawn across that beautiful lustrous eye.

I was not too worried, for by then I had found another nest. It had two eggs, and I built a hide beside it over a week or so. This female was much shyer, and always left before I could photograph her by day. However, I had hopes of being able to watch through the night, certain that the eggs could not be called away like a chick, find out whether both sexes incubated, and other such details. By the time I had the hide ready the nights were dark and moonless. I also had to save the nest from a savanna fire started by some local people. I saw it start from my classroom at the Imperial College, and rushed up to beat it out. It was moving slowly against the wind, so that I managed to stop it before my hide and other equipment were engulfed in flames. However, the people who started it saw me beating it out, and this brought my nightjar-watching to a curious end.

A few days later a detective called on me. What had I been doing with a signalling apparatus on the savanna? It was wartime, and even in Trinidad people were suspicious of such things as flashlights at night. I convinced him with my photographs that I had no evil intent; but I feared the worst. I was right. When I reached the nest, the eggs were crushed by a large Government bootprint, and all my hopes of learning more about nightjars were crushed too. The detective, who rejoiced in the splendid name of Tilbert St. Louis, never really understood why I was so upset about it all.

However, the key to unlock the nightjar door had been put into my hand. I now knew that by slow and careful approach, with a powerful torch, it was possible to creep close enough to a dazzled nightjar to reach out and touch it. I could have caught my Cayenne Nightjar easily enough. Given a good book, I could have taken her home, examined her plumage as carefully as I would a museum skin, identified her, and released her unharmed. Shooting to identify was unnecessary. I would have felt it wicked thereafter to kill a beautiful bird that had given me so much pleasure, so the knowledge was just what I needed.

Later, when I was posted to Nigeria, I put the method to good use. In West Africa there were no less than fifteen species of nightjars, two of them migrants from Europe or North Africa. Of the other twelve, three were relatively easily identified by peculiarities of plumage. These were the Long-tailed Nightjar, the Standard-wing Nightjar, and the Pennant-winged Nightjar. In the last two, the ninth primary in the male is enormously elongated, in the Pennant-winged into a long white trailing pennant, and in the Standard-wing into two long bare shafts, with a racquet-shaped vane at the end. The Pennant-wing was a rare migrant from southern Africa, and the males only began to grow their pennants just before they left Nigeria, about July, on their way back south. The Standard-wings were resident, but outside the breeding season the males moulted their astonishing quills and were then just as difficult to identify as any others, while the females never grew these quills at all.

Freckled Nightjars are among the very few tropical African nightjars which have been studied at all fully. Their attachment to rocky places at least limits the areas that must be searched to locate a nest, in other species most likely to be found only by chance. My friend Peter Steyn watched a pair in Rhodesia, from the time when the nest had a fresh egg to eighteen days after the hatch, when the young died of cold. He took beautiful photographs, evocative to me of the difficulty of seeing this bird at all; she let him take her picture at point-blank range. Not unnaturally, he found that the normal order of things in diurnal birds was reversed. The female sat all day, but was relieved by the male at dusk. Several changes occurred during the night, substantiated by photographs of the sitting bird. The eggs hatched in 18 to 19 days, a little longer than the period recorded for the European Nightjar. The young died during a cold spell when they were eighteen days old, perhaps from starvation, because the parents could not catch insect food. They were only a few days from their first flight. Thus the Freckled Nightjar, whose nest was first found by me in 1944, was, less than thirty years later, perhaps the best known of all African nightjars at the nest. Peter's study was a classic example of the value of observing at a nest, instead of just taking the eggs – which was all I could do at the time.

One might wonder why the male grows such curious appendages; and as one might expect they are used in an extraordinary courtship display. The male Standard-wing flies very slowly, in circles and figures of eight, round the female as she sits on a bare patch of ground. His wings are held forward and downward, and are vibrated through a short arc in a blur, so swiftly beaten that they almost disappear, like the wings of a bee or a humming-bird. The base of the bare shaft of the ninth primary is sharply curved upwards, and when the male is thus flying slowly about, the long shafts extend almost vertically above him. Then the broad racquet-shaped vanes at the tips of the feathers sail slowly above him like a couple of little kites. It is eye-catching; and all the time the male emits a low, intense, churring call, as if deeply charged with emotion. It seems ecstatic, and is lovely to watch.

Other nightjars had no such adornments or plumage that enabled me to identify them easily. However, I soon realized that, even if I could not see a Long-tailed or a Standard-wing Nightjar, I could recognize them by the different tones of their churring voices. Some other nightjars did not call in the reeling or churring tones of these two. Some whistle sweetly; others chuckle; but most churr. One, the White-tailed Nightjar, lived in the grass just outside my house at Ilorin, and nightly came and called on the small lawn outside my front door. By day I used to find them roosting on bare patches of laterite; but try as I might I could not be absolutely certain from the book what species they were. They had a most distinctive call, a repeated, rather melodious, fluting 'Woo-tu-tu-tu' or 'Wha-hu-hu', ending in a series of short 'wut' notes. At length I caught one, with a torch. In the hand, I could examine it so closely that I could even be sure of its racial identity – the Gold Coast White-tailed Nightjar. Thereafter, wherever I went, when I heard that characteristic 'Woo-tu-tu-tu' I knew, without seeing the bird, which nightjar it was.

Gradually, I became more expert in the technique. Driving along at night I often saw dazzled nightjars in the headlights. Then I could stop, get out of the car, turn on the torch with the headlights still bright, and focus on the bird. Switching off the headlights, but keeping the torch absolutely steady, I could then creep up and catch my quarry. As I came slowly closer and closer, the bird would blink in the intense light, but if I did it right it never moved. With the torch held steady in my right hand, some thirty centimetres (a foot) from the nightjar, I could make a quick grab with my left; and again, if I did it right, I had it. I learned that it must never see my head looming above the torch, that I could not move too slowly towards the end, and that on bright moonlit nights the torch did not effectively compete, and the technique would not work. It might have been easier if I had had a wide-mouthed net to slip over the bird. However, with patience, it is perfectly possible to catch a nightjar in one's bare hand; and purists like to do it that way.

Once I held it I just hooded the nightjar with something dark. I never had anything but a sock, which served quite well. The hooded bird would lie immobile in my hand, allowing me to take its wing measurement and note any other distinguishing characters mentioned in the key, until identity was certain. As I always travelled with the books in a special steel box in my car I always had the key handy. Finally, grasping it firmly about the body, I removed the hood and checked any throat or crown markings mentioned. Nightjars are not easily intimidated, and when I removed the hood they usually opened their huge gape, and thrust their heads forward in a snake-like striking motion calculated, no doubt, to deter a possible enemy in nature. Then I could open my hand, and the bird would fly away none the worse. I had another record.

Of course, it was also necessary, when possible, to connect the nightjar with its call. African nightjars emit churring, whistling, or chuckling calls. Most of them churr, but even churring calls vary in pitch, frequency, and tone. Modern tape-recorders, not available to me, have demonstrated that the monotonous 'tuk-tuk-tuk' of the Slender-tailed Nightjar is just a slowed-down churr, and that by playing a churr slower you can make it too sound like 'tuk-tuk-tuk', but maybe much deeper in tone. Likewise, melodious whistle calls are varied in tone; none are quite the same. Evidently, in birds that cannot see each other and communicate by ear, it would not do if they were.

When I was Agricultural Officer in Kabba Province of Nigeria, one of my favourite foot safaris was a place called Abugi, where there were big rice fields. The route passed through a very wild stretch of bush, and I used to break the journey at a place called Gugurugi, where there was a big dome of black rock. Gugurugi was memorable because it was where I first heard a wild Lion roar at close range. I was sleeping outside on a moonlit night, and he sounded off from the base of the rock, not more than a hundred metres away. The magnificent bass voice throbbed on the night like a cathedral organ with all stops out. I rose and went to try and see him. I did not; but since it was moonlight, I heard the continuous calling of an unidentified nightjar. 'Pew-hew', it said, or 'Whow-how', the double note repeated endlessly at intervals of about two seconds, breaking off only when the bird took to flight.

Returning from Abugi ten days later, when it was dark at evening, I determined to identify my nightjar. I climbed the rock with a torch on 23 May 1942 and, as expected, the nightjars called often when they first emerged at dusk, then stopped. I knew the call as one I had heard before round other rocky hills in my district and elsewhere, so I knew it must be a common and widespread bird. When I caught and identified one, I found that it was a Freckled Nightjar. This bird had first been recorded by Boyd Alexander at Gambaga in the Gold Coast in 1905, but the standard reference book recorded only about four others since, from Ubangi-Shari, French Sudan, and Cameroon. There were no Nigerian records. Yet, here it was, evidently very common and widespread in suitable places!

I thought the birds must be courting at Gugurugi in May, but it was not until a later trip, in May 1944, that I could find a nest. I saw the sitting bird, hardly distinguishable from the blackish rock, right out in the burning sun. She was barely visible from only a few metres away. She sat very tight, but when disturbed revealed two eggs unlike any other nightjar's eggs I had seen. Most nightjar's eggs are buff or sandy, mottled with brown. These were greyish-white, spotted with dark grey, nearly matching the stone on which they lay. I felt obliged to take them for science; but they never reached the British Museum and were lost, like so many other things, in wartime.

My torch trick had now led me to a new bird for Nigeria and a nest hitherto unknown to science. Thereafter I needed no torch, only my ears, to tell me that Freckled Nightjars frequented every rocky hill or kopje, but only such hills and kopjes. No rocky kopje or slabs of rock, no Freckled Nightjars; but even quite a small rock slab would harbour a pair, hunting in adjacent savanna. I never found them commoner than they were at Gugurugi.

Nowadays, whenever I camp, I just prick my ears. At dusk, as I sip my first whisky, any nightjar in the area will call; and the calls are all easily recognizable, whether whistles, chuckles, or churrs in different tones and pitch. Then I can indulge in a bit of harmless ornithological one-upmanship with my companions, if any. 'Ha! *tristigma*!' I exclaim. '*Tristigma* what?' they ask. And I explain about the familiar 'Whow-how' that I first identified at Gugurugi long ago. I rather hope that I shall be alone when I hear a new one, for if someone then asks me what it is I shall have to get a torch and go and catch it. I am not as young, supple, quick-handed, and patient as I used to be; and I think I would now descend to using a landing net.

D. TISCHLER

10. Otters

I have always loved seeing and watching an otter, however briefly. I get as much pleasure, and feel it as great a privilege, from seeing an otter as I do from seeing a leopard, though to encounter a leopard at close quarters on foot produces a thrill that an otter cannot. Otters to me epitomize the fluid joy of movement in water. Other creatures such as seals may be more perfectly adapted to a purely aquatic life, but an otter swimming is more graceful than a seal, and it can come out on land and move quite gracefully there too. So they are very special beasts as far as I am concerned.

How anyone can bring himself to kill an otter I do not know. When I was a boy I had a sporran with an otter's mask on its flap – it was the done thing to have such a sporran, and still may be for all I know. Now that I know something of the beauty of otters I would never want such a thing again – after all, a plain leather sporran looks just as well and holds the money as securely. Many otters are killed for their fur, especially the huge species that occur in South America. I would, if I could, prohibit the wearing of otter fur as strictly as I would that of the Leopard, Jaguar, Ocelot, or other spotted cats. And I would prohibit otter-hunting in Britain. I am told that the otter often gets away anyway, but if I have any sympathy for the anti blood-sports league (which I do not because I regard them generally as far-out cranks like extreme vegetarians) it would be on account of otter-hunting. There may be something, if not much, in favour of keeping down foxes in an extremely expensive manner by fox-hunting, though a Scottish gamekeeper does as well without all those trappings. However, the studies that have been done on the otter indicate that it is basically a harmless animal, indeed possibly beneficial in preserved trout waters, since it eats eels and other undesirable fish, because they are sluggish and easier to catch. But decisions on these points, in Britain at any rate, are not made on the basis of scientific evidence and common sense, even today. Otter-hunting may die out naturally through economic causes – it being too expensive to keep a pack of dogs – and I hope sincerely that it will.

Otters occur all over the world, except in Australia, New Zealand, Madagascar, and other oceanic islands, and the Arctic and Antarctic. They belong to the same family as the Stoat, weasels and badgers – the Mustelidae. Yet I do not think that to anyone – even those who kill them needlessly – an otter carries the same air of quiet menace that a Stoat or, more potently, a Honey Badger emanates. They seem lovable or beautiful, rather than vicious.

Their nocturnal habits ensure that they are hard to study, but these may be partially brought about by persecution in Europe, and despite the difficulties they have inspired some notable studies, and at least two classics in the English Language. I read the first of these, *Tarka the Otter* by Henry Williamson, when I lived in Devon myself as a boy; and I know the rivers where he watched his Otters. The desire for arid scientific knowledge is, in the case of otters, nearly always tempered or illuminated by love. Most scientists admit that they find their subjects interesting in the end, but with otters it comes naturally – even when they bite you.

Otters live in all habitats from the mountain streams we think natural to them in Britain to steaming tropical jungles, where great rivers thread through swampy floodplains. The smallest species, the East Indian Clawless Otter is about a metre (three feet) long, but the largest, the Giant Brazilian Otter, can be two and a half metres (nearly eight feet) long, one of the largest and most powerful animals in its family. I have never seen one, but those who have tell me that they are just as delightful to watch as any other otter. Certainly, one does not think of them as possibly dangerous, though they are the size of a small leopard, weighing up to 36 kilogrammes (80 pounds). True, they do not have the awful retractile claws of the leopard (and I know what those feel like), but it would be no light matter to be bitten by one.

All but one species of otter are basically freshwater animals, though in the Highlands of Scotland the Otter swims freely along the shores and around islands. The Sea Otter, which lives in the Aleutian Islands and off the shores of California, is a special beast – a tool-user. I count among my most glorious memories a lovely sunlit day spent at Point Lobos near Monterey, where we watched the Sea Otters playing in the surf. They dived to collect abalones from the bottom among the kelp, surfaced, and lying calmly on their backs, unable to sink, they bashed the mollusc against a stone held against their chests till it broke and they could eat it easily.

Sea Otters were once almost exterminated in California for their fur but now they are recovering and spreading. An animal spending its time in water must have a waterproof pelt. That of otters is soft, silky, lustrous. I can see why women would want to wear it; but I cannot forgive them for endangering beautiful and lovable animals to do so.

I saw my first otter in a dramatic place, beneath the gigantic cliffs of Hoy, in Orkney. In those days I was studying and photographing Great Skuas and other birds, just after completing my Honours year at St.

Andrews University. On Hoy between the 137-metre (450-foot) pillar of the Old Man and the Kame, or Comb of Hoy, which rises to 365 metres (1,200 feet), there is a range of magnificent sandstone precipices, at two points – the Sow and the Carl – about three hundred metres (a thousand feet) sheer, or overhanging. I took a childish joy in lifting a big stone and hurling it over, leaning out to watch it go down and down, twirling as it fell, to hit the sea or the rocks below with a crack that set the seabirds on the move. The bigger the stone the better on a cliff of that height; but one must have good balance and a nerveless head for heights.

I was about to do this from one particularly satisfying place I had found, where the cliff overhung to a degree that would make the rock hit the water some twelve metres (forty feet) out from the base. In order to make certain I did not kill some unfortunate lobster fisherman I looked over first, and at once saw a brown, slug-like object on one of the big square boulders fringing the cliff-foot. I could not make out what it might be until I put the glasses on it, when I saw it was an otter. I lay on my belly and watched it over the edge for a few minutes; then it disappeared into the spaces between the boulders where it was doubtless hunting crabs.

The next otters I saw were a family of parents and two cubs on a forested stream in the mountains of Trinidad. We used to use such streams as highways into the dense rain forest and, in the heat of the afternoon, cool from a bathe in the clear water, used to lie on a convenient log spanning the stream and whistle, in a certain way, to call up forest birds. We were doing this one day when the otters swam downstream and into the neck of the clear pool above us, about eighteen metres (twenty yards) away. We could see their heads – two big, two smaller – and parts of their lithe bodies under water. But they saw us the instant we moved, and so we lost the chance of watching them swim under us at a range of a metre or so. They turned and retreated at once upstream. What they ate there I know not; perhaps crabs, as to my recollection there were no sizeable fish.

Nigerian otters, of which I saw many, certainly ate crabs and freshwater mussels. The common species here, the Clawless Otter, *Aonyx capensis*, is a very big one, a male being about a metre (four feet) long and quite heavy. Since they do not regularly catch slippery fish, but crabs and molluscs, they have no need for sharp claws, and their front paws are soft, like little webbed hands. I know, because I have felt them grasping my own. When the rivers were low in the dry season I quite often saw these otters sitting on a sandbank or a rock, sometimes upright, as a squirrel sits, holding a crab or a mussel in their front paws and taking crunching bites out of them as you or I would bite into a slice of water-melon. These ground-up remains of crab and mollusc shells remain as spraints, laid on boulders in midstream as the otter marks its territory. I have never found a fish scale in such spraints; and I know that these otters in Kenya trout streams seldom if ever caught healthy trout.

Gavin Maxwell had not, at that time, written his classic *Ring of Bright Water* which has done more to inspire public sympathy for otters than anything else could have done; but his second tame otter was one of these African Clawless Otters. I too had a pet otter for a short time. I acquired it when I was on a long and dreary launch trip in the stifling hot, almost unmitigatedly vile mangrove creeks of the Niger Delta. Some fishermen had caught and killed the mother to eat, and my attention was attracted by the piercing cries of the still living small pup which had been with her. I do not know what they intended to do with it, but I bought it from them for a shilling, more than its value to them as meat.

Asaberu, as my Hausa henchman Momo immediately named the little creature, at once presented a problem. He (we ascertained his sex as soon as we handled him) was hungry and he signified his need with a piercing whistle impossible to ignore. In those days we used tinned condensed milk; and as the pup was still very young, his eyes dim and baby-blue, milk was what he wanted. I had no bottle, of course, but necessity is the mother of invention and I contrived something out of a small syringe and the twisted top of a tin of canned fruit to make a funnel and a dropper, with which I fed him, drip by drip, until his whistles ceased and he lolled sideways to sleep.

Not for long, however! Twice a night, and three times by day, we went through this performance. He soon became more practised at it and did not spill so much. My ration of tinned milk was limited, so I diluted it and added sugar. I did not have any idea about a baby otter's diet, but fortunately he throve. In the night his piercing screams woke me from tossing fretful sleep, all that was possible in that miasmic, humid, malaria-stricken region, and I used to warm his milk to blood-heat over my bush-lamp. By day he slept; and ten days later we safely returned to Umuahia, where I was at that time stationed.

I now had two pets, an otter and a half-grown Leopard, which had been forced upon me by my senior officer, to whom it had been presented, but which he emphatically did not want. The otter was adorable, the Leopard dangerous. As big as a retriever, it was incredibly strong and was kept chained to an overhead wire in an enclosure. It never even looked like becoming really tame, but it used to enable us, from time to time, to put on an entertaining show for people who came to drinks. It was looked after and fed by an Ibo boy who had few other virtues except that he was unafraid of the Leopard. While we were sitting there, he used to come proudly in, bearing a dish of bones, soup, and rice, and announce solemnly, 'Leppaahd chop Sah.' The Leopard would then be released and, growling, commence his meal. We waited for the climax, having seated the chosen victim of our little joke in a certain chair. When the Leopard had finished it either walked or sometimes, best of all, leapt some three metres (ten feet) into that chair, where it proceeded to knead the thin palm-beach trousers of the unfortunate victim with its frightful 25-millimetre (one-inch) claws just as any cat making itself comfortable on anyone's lap will. As a result of protests we cut its claws later. It took five strong men to hold it, one to each foot and one to the head; and it spent much of the night debarking a tree in its enclosure and had the claws quite sharp enough again next morning. I had visions of rearing the creature and releasing it in the wild, but eventually saner counsels prevailed, and I shot it soon after I was transferred to a place where continuous touring would have made it impossible for me to care for it decently.

Asaberu was different. The day after returning to Umuahia I went down to the U.A.C. store and bought a baby's bottle and a nipple. I filled it with warm milk, and he at once grasped what it was for. Seizing it in his little soft hands, he sucked and sucked, to the accompaniment of a series of satisfied cries, 'Yum-yum-yum.' He was the only animal, not excluding my own infant son, I ever heard utter anything really like that sound, considered generally expressive of gustatory satisfaction. He sucked and sucked and his belly swelled visibly. Then, when he could hold no more, he fell on his side with a final satisfied cry, and was almost instantly fast asleep.

He throve, and was a delight. His exquisitely silky-soft pale grey fur began to darken, he doubled his weight, and his eyes lost their baby-blue and began to see things a bit better. To handle him, to place him on one's shoulder and feel his silky pelt brushing one's neck and ears as he snouted around, was a sensual pleasure. Then, when I had had him a month, I was transferred back to Okene, a place I loved and was glad to return to, after Umuahia. There I could lead the bush life of safari I wanted to, including trips on the Niger and Benue in Sanders-of-the-River style. But the animals were a problem. No lorry could be spared immediately, so I somehow had to cram everything, Leopard, Asaberu, servants, and all, into my 10-hundredweight Chevrolet van, and make the 480-kilometre (300-mile) journey in a day.

Asaberu survived the trip well enough, mostly on Momo's lap in the front. He had to be bottle-fed at lunch-time, but a really heavy meal then left him quiet till evening. The Leopard we put in a dark hut to settle him; but one look at those glaring green eyes next morning decided me that it was no use. Even the lion-hearted Momo, the finest African I ever knew, was afraid of him. So I got the gun and shot him at once.

Soon after I had to go and visit an outstation, and rather than take Asaberu on another long journey I left him with Momo, who knew what to do. Either the long journey from Umuahia had upset his little interior, or a too early attempt to add some protein solids to his diet to augment his now voracious need for milk did not suit him. When I returned he was dead; and Momo's graphic description of how he had suffered and died did not help. I had hoped to rear him, teaching him to swim in the Okene reservoir, and then, when he was big enough and could find his own crabs and mussels, release him in a big river in wild forest where there were plenty of his kind. We buried him in my garden; and he remains a sweet and poignant memory. I ought to have done better, but I knew nothing about rearing otters and in my ignorance did the best I could. I gave him a chance.

Thereafter I have seen otters from time to time, but only for brief moments as a rule. Oddly, I have never seen one in Kenya trout streams, though I have often waited for the late evening dry-fly rise, when they might be on the move; and I have found their paths beside waterfalls, where they must emerge on to land to circumvent the obstacle. Even on lonely islands in Lake Victoria, where I used to study Fish Eagles and sometimes camped for a night or two, I never saw one, though my brother once came on a family, quite unafraid of him, on an island off Karungu Bay which local fishermen avoid as a place of bad omen.

Fishing for sea trout at night in Devon and elsewhere I have had brief encounters with otters. Once a big dog swam straight down the middle of a weir pool near Holne Park on the Dart. He did not see me until he was level with me, ten metres away; but he then dived and slipped over the weir under water, making only the faintest ripple on the glassy glissade of barely ankle-deep water pouring over the lip. He must have flattened his body till nothing appeared above the surface. Another time, when I was sitting on the edge of the salt-marsh waiting for the tide to back out of a sea pool, an otter swam up the pool and came out on the shingle right at my feet, not more than two metres away, peering this way and that at this curious object outlined against the sky. At length he decided he did not like what he saw, gave a disapproving 'wuff', and returned to the pool. These moments demonstrated that at night an otter has poor sight when out of water, probably not even as good as mine. Also, that the recent passage of an otter through a pool full of sea trout has no effect on the fishing after a few minutes for in both cases I caught or hooked good sea trout within a quarter of an hour of their passage. So otters need not be persecuted on that account.

The last otter I saw was on the Kirkaig River in Sutherland. I was fishing the falls pool, on a day of soaking wet, when the spray from the fall in spate and the rain combined to produce that state of abject cold, soaking misery that only a keen fisherman will endure for long. The pool was full of salmon, scores of them, and I fished from a rock two metres above four and a half metres (fifteen feet) of deep black peaty water. As my fly passed below me after an upstream cast I saw a huge whitish shape rising from below. 'By God, that's some fish,' I thought, dreaming of a fifty-pounder (23 kilogrammes). Then: 'What on earth is a seal doing here, ten miles upstream?' And finally the head of a huge dog otter broke surface, eyed me with black beady eyes for a moment, and sank below, never to reappear.

I have even, as an unsuccessful fisherman one evening, been indebted to an otter. I was fishing the sea pools of the Dionard River near the Kyle of Durness, on the falling tide, and I rested briefly on a flood-stranded tussock of soft green grass – the obvious seat from which to watch the tail of a long pool. Nothing stirred there, and I went on further down. Coming back empty-handed an hour later I found on that tussock the silver body of a freshly killed sea trout $2\frac{1}{2}$ pounds (1·15 kilogrammes) in weight. An otter had killed it, eaten the soft flesh of belly and the guts, and decently left all the best of the sea trout for me. I am not disdainful of such favours and I fed gluttonously on that sea trout in my motor caravan the very next night.

None of these encounters taught me much more about otters than I already knew, but the sudden blinding light of revelation into the ways of otters came on the same trip in Sutherland when the otter came out of the water to my feet and another left me that opportune sea trout to eat. In the first ten days of June that year (1967) we had a spell of that glorious weather which is so rare, and therefore so doubly welcome, in the north of Scotland. Day after day the sun shone from a sky of unclouded blue; the sea of the Pentland Firth was flat calm except in the tide rips; the rivers shrank to the curses of assembled fishermen; and even the sphagnum bogs dried till it was almost a pleasure to walk over their crisply springy surface. Best of all, the broom and the whins burst into their full glory of golden bloom, which we call the 'Field of the Cloth of Gold'. It was a time to savour and remember, a rare week of sheer delight.

But it was too hot for strenuous eagle surveys, which is what I was doing, and I allowed myself one or two relatively easy days. On one of these I had climbed from the road along the east bank of Loch Hope up the valley of a stream with a waterfall, where I later bathed on my way back. I reached a little cliff above a small loch, Loch na Seilg, and there lay in blazing sunshine, tempered by a gentle breeze, looking down thirty metres or so (a hundred feet) into the crystal-clear waters. I could see the weeds on the bottom, even a half-pound (225-gramme) trout cruising slowly about and taking an occasional insect below the surface. I lay with my head on my arms, comfortable and content, revelling in the warmth and a sense of physical well-being, for I was fit and tough after two weeks on the hill.

I was looking straight at the point where the burn left the loch, and in the glassy-smooth water there I saw the vee of a bow wave approaching. Thinking that it was probably a merganser fishing with its head and neck below water, as they do, I glassed it. No snaky head and neck appeared, however, and a glance assured me it was an otter, slowly swimming straight towards me and diving at intervals. He swam on the surface with his back showing, but his head submerged. He was in fact, doing what I often do on the Kenya coast, viewing fish under water with only his head submerged. He had to raise his head to breathe from time to time, for he had no mask or snorkel; but he managed perfectly well.

On and on he came, straight for my rock, and as he (or maybe she) neared me so I could see, each time he dived, a little better what he was doing. At length he was cruising round in the bay below me, where I could see every detail of his movements. He dived, and using his hind feet to propel his graceful body without effort, swam about just above the weeds, not far from the bottom in three or four metres (ten or twelve feet) of water. He caught nothing; the half-pound trout had wisely made himself scarce. The otter's body was silvered with air bubbles caught in his pelt, and these floated to the surface in the chain that betrays the underwater movements of the otter to the master and followers of otter hounds that would kill him. I could see his head moving from side to side as he searched for prey. Then he finally surfaced, almost filling the field of my 12-power binoculars, and swam quickly to shore to disappear among the talus of big scree at the loch edge below my cliff. I never saw him reappear, and assumed that there was a cool holt there where the hot day would be spent.

He never saw me from first to last, but in those fifteen to twenty minutes I learned more about otters than in all the other encounters combined; I shall probably never see another so well, or for so long, or so revealingly. I know now that an otter has poor sight in the semi-dark of a Scottish summer's night and that it hunts under water by (at least sometimes) swimming on the surface with head submerged searching for fish, then diving to pursue among weeds where they may be hiding, hard to see. My only regret is that the slow, leisurely, ineffably graceful movement of the questing otter did not break into lightning pursuit of that trout I knew must be somewhere about. Then I should have seen him come out on a rock and eat his trout, and I should not have made him drop it for me.

Since then I have never seen another otter except for brief moments and I have wondered exactly what adaptations enable an otter to see thus under water. I, snorkelling on the Kenya coast, see perfectly under water with a mask which separates the water from my eyes. The merganser duck and, I suspect, the cormorant and other underwater snoopers such as the darter, have nictitating membranes which they can draw over the eyeball and which are transparent, so that they can see their prey. The standard books on mammals and on otters (by C. J. Harris in the World Naturalist series, published by Weidenfeld and Nicolson) do not greatly enlighten me about otters. All they say is that otters have fair vision under water, perhaps better than above water. I have not found (and I admit too that I have not very seriously looked for) a detailed anatomical study of an otter's eye. And I am not going to kill an otter and examine its eyes to find out for myself, though some of those who do kill otters for their pelts might think of doing such a thing as a useful by-product of their otherwise useless cruelty.

Whether I shall ever learn how it is that an otter can see well under water when his sight is poor, if adequate, to get about on land, I do not know. But I can be glad that fate took over on that little cliff above Loch na Seilg on a hot Scottish summer day. I would have had a good day anyway for I later climbed Ben Hope in a clear lambent evening and satisfied myself, duty bound, that there was no eagle's nest in the northern corrie – as had been reported. But the otter made it a day in a million, never to be forgotten, a bottle of the finest, purest, amber-coloured, Scottish-hill-water wine.

D. TISCHLER

11. Pelican Island

All pelicans are fascinating birds. Combining a somewhat grotesque appearance with a certain grave dignity, their ponderous waddling bulk on the ground becomes a soaring marvel of grace and beauty in the air. The beak, which is really what makes them look grotesque, is actually a wonderful multi-purpose tool. It is not only capable of catching the fish on which they live, but is used as a signal in nuptial display, and is capable of almost incredibly delicate adjustment when feeding the tiny, new-hatched chick, one-fiftieth of the parent's bulk. It may not be able – as the rhyme says – to hold enough food for a week, but can certainly hold more than his belly can. The elastic pouch – itself a wonderful adaptation – might hold about five and a half kilogrammes (twelve pounds) of fish if packed tightly, like trout in a creel. However, no pelican, with such a bag, could go far. If it took off it would nose-dive ignominiously, as it would be over-weighted in front. As an escape reaction, a gorged pelican will regurgitate all the fish it has just laboriously caught, to aid a swifter take-off.

There are seven or eight species of pelicans in the world, varying from the enormous Great White Pelican, *Pelecanus onocrotalus*, to the much smaller but still very large American Brown Pelican, *P. occidentalis*. The living species are but a remnant of those that once existed, for millions of years ago pelicans flourished in areas of warm shallow seas. Nowadays most species of pelicans are reduced, some, in North America or Europe, almost to danger point, while others are still quite abundant. Some are especially sensitive to poisoning by organo-chlorine insecticides, and they cannot stand encroaching civilization. For these reasons, those that are not yet threatened, and apparently still flourishing, may easily be threatened with extinction before the end of this century, if we do not look after them.

Pelicans are divided into three main groups by their fishing and nesting habits. The first of these, including the American Brown Pelican and its western race, the Chilean Pelican, are mainly marine or estuarine, and do not frequent inland lakes. They are distinguished from all others by the fact that they dive for fish. Flying low over the water, or soaring against a breeze, they plunge with the neck retracted when they sight a fish. Slow-motion films have revealed that, just as the pelican approaches the surface, it shoots out its neck and beak, piercing the water well ahead of the body, and enabling it to catch a fish a metre or so under water. No other pelican can do this; and it is a grand sight to see a soaring flock of these great birds diving, one after the other, into a shoal of fish.

Brown Pelicans breed sometimes in trees, sometimes on the ground. Two other rather larger species, the Pink-backed Pelican of Africa (*P. rufescens*) and the Spotted-billed Pelican of Asia (*P. philippensis*), both breed in trees and apparently form another species pair. I do not know exactly how the Spotted-billed Pelican fishes, but the Pink-backed Pelican normally catches its prey by stealthily swimming about, and lunging with extended neck and beak to catch its quarry. One may see Pink-backed Pelicans moving very slowly about in lagoons covered with water-lilies, their heads retracted on to their backs, beaks laid flat on the chest. They adopt, in fact, a low profile. Stealthily they approach a fish, and then the beak is suddenly shot forward. If skilful or lucky the pelican has got it; if not, it adopts its low profile again and swims slowly on.

The other four species, the Dalmatian (*P. crispus*), the Great White, the American Great White (*P. erythrorynchus*), and the Australian (*P. conspicillatus*), are all much larger than those mentioned so far, the largest, the majestic Great White, being one of the largest flying birds. A big male Great White can span 3 metres (10 feet) and weigh over 20 pounds (9 kilogrammes). All, as far as I know, breed in huge colonies on the ground, on inaccessible islands or rocky hills, never in trees. All fish, as a rule, by what may be called the 'scare-line technique' which is also used by many human fishermen. This depends on the reluctance or fear of certain fish to swim under or past an obstruction or shadow.

A group of 8 to 20 pelicans, averaging about 10 to 12, gather together and swim forward in a horseshoe formation, open end foremost. So here we have a curved, nearly circular shadow and an obstacle composed of swimming webbed feet and bodies. The fish swim away continually towards the open end, so that more and more collect inside the moving horseshoe. Then, as if at a given signal, every pelican flips his wings upwards and part-open, suddenly intensifying the shadow. At the same time all dip their beaks together into the centre of the horseshoe. Some withdraw their bills empty; others can be seen swallowing a fish they have caught. In effect, it is as if a dozen landing nets were dipped simultaneously into a shoal of fish; some are almost bound to be netted. A big fish can be seen sliding down the thin, upstretched neck of the successful pelican. The group then swims on, repeating their skilled netsmanship at intervals, sometimes joined by others, until all are satisfied.

If we now take a closer look at that marvellous, if slightly grotesque beak, we can see how it works as a multi-purpose tool. The upper mandible which, in a big male Great White, is nearly 46 centimetres ($1\frac{1}{2}$ feet) long, is stiff and strong, with a central rib, serrated on the inner side, which will prevent a caught fish from slipping forward. Two parallel channels run down either side of the rib; and at the tip there is a hard, nail-like structure which, when upside down is hollow like a tiny cup. The purpose of this will become apparent later. The lower mandible is quite different. It consists of two thin flexible bones below which hangs the pouch, a strongly muscular and highly elastic organ capable of very wide expansion. The pouch and lower mandible are normally tucked away under the stiff upper mandible, but when the pelican plunges its beak into water to catch a fish, the slender flexible bones expand rapidly into a broad spade shape, and the pouch becomes an efficient scoop, acting like a landing net. The fish caught, the pelican retracts its neck, the water drains out, the flexible bones again straighten and fit under the upper mandible, and the wriggling fish is held in the now shrunken pouch. I have never yet seen a fish escape the pouch once inside it, though doubtless it happens to pelicans as well as to me. The pouch itself is often brilliantly coloured, especially in the breeding season, and is then used in display, to other males or to females. So every part of that apparently grotesque schnozzle is actually beautifully adapted for some purpose, a marvel of evolution indeed.

The first pelicans I ever saw were the Brown Pelicans of the West Indies. They bred on Saut d'eau Island, off the north coast of Trinidad, and I studied them to some extent, mainly as a bird photographer. In those days I was not a very knowledgeable ornithologist; but looking back, I can now see that I did note many of the features common to other pelicans. Without particular wonder I observed the long and apparently elastic breeding season – with nearly mature young and fresh eggs, several months apart, in nests at the same time. I noted also that relatively few pairs – about 150 to 200 – actually bred out of a population numbering many thousands round Trinidad. I observed the obviously poor breeding success, with many young falling out of nests, and many nests failing altogether to reproduce their kind, and various other things which seem the common lot of pelicans. They all seem to be irregular and opportunistic breeders, and if I had done my elementary arithmetic, I should even then have seen that the Methuselaic mien of a big pelican is not misleading. Breeding irregularly and unsuccessfully, they must be long lived to survive at all.

Both Pink-backed and Great White Pelicans are common and widespread in tropical Africa. There are possibly 100,000 to 200,000 pairs of

Great White Pelicans which, I suspect, outnumber their smaller, more adaptable, less gregarious cousins. I first saw Pink-backed Pelicans along the Niger floodplain (where they enjoy the splendidly evocative Hausa name *Kwasa-kwasa*), but they were not very common and did not breed there. Friends living farther north told me of easily accessible colonies in big trees in villages, but in wartime I could not reach these. In Nigeria there was also one suspected colony of Great White Pelicans on an almost unclimbable block of rock called Wase. It certainly existed until recently when mountaineers, in their selfish determination to climb rocks no matter how or what other damage they may do, disturbed it. I never could reach Wase either.

When I became Provincial Agricultural Officer in Nyanza Province, Kenya, I came to know of a big breeding colony of Pink-backed Pelicans at Oyugis in south Nyanza. This was situated 27 kilometres (17 miles) from Lake Victoria in several large trees growing at the head of a little valley in cultivated country. Why they bred there and not in innumerable more easily accessible fig-trees along the lake shore is a mystery to this day. The breeding trees were a remnant of forest that once filled the valley. They were defended not only by swarms of extremely aggressive bees nesting in hollows in the trunk, but also, rather refreshingly, by local people. On my first visit a Luo elder came rushing up to me, demanding what I was doing, and roaring at me that I must on no account harm the birds, which his people thought of as sacred. I assured him that I had no intention whatever of harming the birds, but only wanted to look at them and watch them. So it was agreed that I could climb one of the smaller trees, a fig, also infested with bees, in which there were only a few nests. The main nesting tree, which has since collapsed, was unclimbable; and the bees would sting me even if I ventured to within thirty metres of the tree.

I had too much work to do to make other than occasional visits to this colony. I was also, at the time, fully engaged on flamingo research, and had to keep my eye on that ball in my limited spare time. However, in the few visits I was able to make, I established broadly the limits of the breeding season August to March, and by sheer luck recorded the incubation period in one nest fairly accurately. It had one egg when I first climbed it and the first chick of two had just hatched 33–4 days later. Zoo-keepers tell me that in captivity all pelicans hatch in about 30 days, but all the wild ones I have observed take rather longer.

Later, a friend of mine, Victor Burke, studied this colony more intensively, and we wrote a joint paper about it published in *Ibis* in 1970. Although far from a complete study, it is the best I know covering the breeding of Pink-backed Pelicans. It would not be difficult for any competent scientists to do better, either with the Pink-backed Pelican or the similar Spotted-billed Pelican, for they are both faithful to certain special trees or groups of trees, sometimes quite far from water. One could erect a fine platform hide in the tree outside the breeding season, and spend all the time one liked up there. Pink-backed Pelicans do not seem to desert their nests regularly or easily – perhaps because they breed in trees – and both Victor and I found that we could repeatedly climb to the same nest, or to a hide close to a group, without causing desertion or much anxiety. Thus the Pink-backed Pelicans failed to warn me of the extreme nervousness of their bigger relatives, the Great White Pelicans, when I came to study them.

Great White Pelicans, the most majestic of all the species, became my constant companions when I was studying flamingos on the alkaline lakes of the Rift Valley. I used to see their squadrons flying high along the Rift by day, magnificent in their regular vee formation against the blue sky and clouds. At the north end of Lake Hannington (now Lake Bogoria) flight after flight, sometimes over a thousand strong, used to descend in the evening to the mud-flats near Luboi. Zigzagging down from a great height, they gave a wonderful display of control and dexterity in the air, before alighting with a thumping of great wings, gilded by the evening sun. Next morning when the sun grew warm they took off, and usually flew along the west side of the lake to an area with many geysers, where rising columns of warm air gave them the lift they needed to surmount the 600-metre (2,000-foot) escarpment between them and Lake Nakuru, or other lakes to the south. In those days there were no fish in Lake Nakuru, and the pelicans could only feed at Lake Rudolf, Lake Baringo (which they seldom frequented), Lake Naivasha, and thereafter Lake Manyara. Since they often alighted for the night on lakes without any fish, they could evidently fly for some days without any food at all.

Up to 1961 there were no recorded breeding colonies of Great White Pelicans in Kenya or north Tanzania. A vast breeding colony, numbering perhaps 40,000 pairs, was found in the inaccessible swamps of Lake Rukwa by the staff of the Red Locust Control, headed by the late and much loved Desmond Vesey Fitzgerald. To this day we do not know if this colony is regular, or how many breed yearly if it is. It seemed too far away for me to reach from Nairobi in my very limited spare time. Since it was situated in deep swamps it could only be easily overlooked from the air. To reach it on the ground, even for one day, would have been a difficult business.

Then, in 1961, came the heaviest rains ever recorded in East Africa.

They resulted in widespread flooding, since they followed a severe drought, with intense run-off from nearly overgrazed bare ground. Lake Natron, normally a nearly dry expanse of horrid crystalline soda plates, with whose lethal character I was only too well acquainted, filled up with water about a metre deep. At the south end there was a group of large rocky spikes, normally protruding from bare expanses of dry brown mud. Now they were surrounded by quite deep water. The Great White Pelicans, opportunists that they are, bred on them in great numbers. It was the sort of thing that could only happen once or twice in a century, if that.

I heard a rumour about this colony and on 9 August 1962 flew over it with Charles Guggisberg. We had earlier looked at the huge 1962 colony of flamingos on Lake Magadi, and we only had an incidental look at Lake Natron. However, there they were, covering at least fifteen islands, possibly 10,000 pairs or more. I flew towards them and began to descend to try to take photographs and make counts. However, it was 10.30 in the morning, and the air was full of pelicans. I very soon realized that I was flying into acute danger, for had our frail Piper Colt aircraft collided with only one pelican we should almost certainly have crashed. So I withdrew, with my heart in my mouth till we were clear of the soaring squadrons. If Charles Guggisberg was as frightened as I was he did not show it.

I felt I must reach this colony somehow. Early in September, when they should have been hatching, I had a try. I could only spare a long week-end, but I thought that by using the light boat I had used to approach flamingo colonies on Lake Elmenteita a few years before, I ought to be able to manage. I drove down the track to the magnesite mine at the foot of Mount Gelai, which I had traversed in such agony a few years earlier. It was now much better marked, and went on beyond to some large hot springs. In theory, this track continued to the south end of Lake Natron. However, the rains had cut out a deep gully with vertical banks, and beyond it there were large scattered boulders, so that without a gang of men with spades and pickaxes I could get no further along the track.

I went down to the shore and had a look. I realized at once that in my light but unwieldy boat, easily blown about by the wind, I had not a hope of reaching the islands in the time I had available. It would have been different if I could have driven to my planned point 16 kilometres (10 miles) nearer, but from here I should have had to row that distance each way. I knew that strong winds would rise, and could prevent my return completely, probably stranding me on inhospitable mud-flats without any fresh water, in scorching heat. So I thought better of it, and instead walked briskly for about ten kilometres along the shore until I was within easy binocular range of the colony.

It was now midday, and scorching hot. At the same time a few weeks earlier the air had been full of circling soaring pelicans, and they should have been still more obvious now, with young to feed. Even in the heat haze I should have seen them clearly, for the breeding rocks were now not more than eight kilometres away. The squadrons of soaring pelicans would have sparkled in the sun as they circled to gain height, or woven arabesques in swift descent. There were none. They had gone. All, or nearly all, of this huge colony had left eggs and young, and abandoned their attempt at breeding altogether. Since then I have seen similar sudden desertion twice at Lake Elmenteita, so I believe I was right.

The reason did not seem far to seek. The whole shoreline of Lake Natron was rimmed with a tideline of small dead fish, Tilapia, each about 75 to 100 millimetres (3 to 4 inches) long, piled in a solid strip about a metre wide and centrally 76 millimetres deep. It did not occur to me to sample the numbers per square metre – they were partly dried and stinking – but evidently a colossal potential source of food had suddenly perished all together. Such mass die-offs of small fish had been recorded from Lake Magadi also. It seemed obvious that such a sudden loss of food supply, which may also have happened at Lake Manyara further south, could easily account for the desertion of the whole pelican colony. Allowing about 1,000 little fish per square metre, and a continuous strip 240 kilometres (150 miles) long and averaging a metre wide all round the lake, there could have been 250–300 million dead fish, weighing perhaps 8,000 tonnes. There might well have been twice that amount.

Anyway, the pelicans were gone; the chance was lost. I could only rail at fate. I did a few other odd things, and found, near the fresher spring where I drank in 1954, how a male lion had stalked a group of resting pelicans behind a tuft of reeds, rushed out, slipping on the mud, but caught one before it could escape, and eaten it. I wished I had seen it happen; but the tracks and the mass of feathers where the lion had plucked his prey told me all I needed. However, the longed-for chance to spend even a few hours in a colony of Great White Pelicans in full swing was lost. It would, I knew, have been a red-letter day in life for me, and as I drove home I was disappointed that this great day had been snatched from me. Still, somehow I felt that my chance would come.

It did, four years later, when I was in Ethiopia as a member of a two-man UNESCO team to develop wildlife conservation, with a chain of National Parks and game reserves. I was nearing the end of this assignment when Ato Berhanu Tessema, then head of the newly established

Department of Wildlife Conservation (and since, I fear, either exiled, imprisoned, or murdered by the gang of Marxist assassins that governs that poor benighted country) told me with some excitement that he had found a colony of pelicans on Lake Shala. He and a friend, Ato Aseffa, Vice-Minister of the Interior, had visited an island in a boat and had found it covered with pelicans. He did not know what species, but had taken photographs and these showed that the downy young were *black* – therefore Great White Pelicans, not Pink-backed. Excitement poured through me like a slug of the best whisky!

Brian Wood, then head of the Biology Department at the University of Addis Ababa, was studying the chemical content of the waters of the Rift Valley lakes in Ethiopia. It was not too difficult to twist his arm to take some samples from the depths of Lake Shala, using the boat provided for his work. We could visit the Pelican Island at the same time. So we did, on 21 March 1965. As we approached, I could see the white bodies of the nesting pelicans massed on a gentle slope, and hundreds of others in the air. We circled the rock and landed on a little shingle beach, out of sight of the main colonies. Cameras at the ready, we crept up to the rocky ridge in the centre of the island and looked over. There they were, some less than twenty metres from us. I began taking photographs, but sadly had only black-and-white film in my camera.

Almost at once it struck me, too, that these great birds, normally phlegmatic and pompous-looking, were extremely nervous and ready to desert their eggs. Some, with small chicks, remained staunch, and let me photograph them at close range. Others, with eggs, waddled slowly away, looking nervously over their shoulders. As soon as the eggs were exposed Egyptian Vultures, waiting on dead trees for their chance, swooped down and began to peck at the eggs. Some they picked up and hurled against rocks and stones till they broke open. I realized that I was causing unexpected damage, and retreated at once. I had only been there about a quarter of an hour, so we crept to a more distant viewpoint and stayed watching them from there without disturbing them further. It was my first really good view of a longed-for sight, and even in that short time it was as spectacular and rewarding as I had always thought it would be. For the time being I had done what I could, and my cup was full.

Part of the wildlife plan I had designed for Ethiopia was for three months' research on the Mountain Nyala, about which I have written in Chapter 8. In 1965 Emil Urban, the best friend I ever had, had been posted to Ethiopia as a lecturer in the Biology Department. I reckoned that after a month living high in the Bale Mountains I would need a break at lower altitudes; and what better break could I have than a few days studying the pelicans on the island in Lake Shala.

Emil had borrowed a cockle-shell aluminium dinghy from a friend in Ethiopian Airlines. It was supposed to be unsinkable, but he always regarded it as pretty dubious, especially when loaded. I never told him so, but my personal view was that it would sink like a stone. With this frail but easily transportable craft we went down to Lake Shala on 22 January 1966. We carried the boat on top of Emil's Volkswagen, and in the back we had the outboard engine and our food. We thought the weather would be dry in January, so we had taken no tent. We launched the boat at about 4.30, and set off. We had thought the trip would take only an hour or so but there was a considerable sea running, and our boat was loaded to the gunwale, so we actually did not land on the island till towards dusk. Sometimes we only had a few centimetres freeboard.

From my visit the previous year we had estimated that in January we should find the pelicans with eggs. However, we had no very clear idea of what we would see. As we approached the island we could see that most of the gently sloping side was covered with pelicans, and it looked as if some would be very close to the rocky ridge from which we could watch most easily.

We hurriedly got all our gear ashore. We had no time to go and look at the colony for, unexpectedly, a thunderstorm was brewing to the east. Since we had no tent, all we could do was to put our gear under a tarpaulin (a nice new one, fortunately) and weight it down with stones. For a while I hoped that the storm would not reach us, and sat on a rock watching it approach. Like all approaching thunderstorms which one can see no way of avoiding, it was at the same time awesome and menacing, yet spectacular and beautiful. As I sat there watching the grandeur of the elements I saw, crossing the van of the thunder-cloud, a strange white spinning spiral, like the threads in the inside of a glass marble. I thought at first it must be some kind of waterspout, but when I looked at it with binoculars I saw that it was a column of avocets of all things, several thousand strong and perhaps a hundred metres or so high, moving rapidly from south to north, perhaps to escape the storm. The spinning spiral effect was produced by the light flashing from their pied wings as different parts of the column turned this way and that, as flocks of waders will. To this day I think of it as one of the most extraordinary things I have ever seen any birds do. To see it thus, crossing the van of a fearful thunder-cloud, was an amazing sight.

We still hoped, rather forlornly, that we might escape a wetting. However, when the first drops began to fall there was nothing for it but to retreat beneath the tarpaulin, taking care not to dislodge the stones, and hope for the best. We had a few short bits of stick with which we could raise the centre of the tarpaulin and so allow some of the water to drain off. These we gingerly inserted and raised, till we had about forty centimetres (fifteen inches) of headroom. Then we found that, in our hurry, we had not blown up our air-beds. The ground was exceedingly hard and stony, barely mitigated by the dense growth of star grass covering it and concealing the stones. Blow up those air-beds we must; and somehow we did it while actually lying on them. This required a great deal of extra lung-power, but also meant that we could inflate them till we could no longer feel the stones, and then stop. The rain lashed down, but thanks to the fact that the tarpaulin was new and we had those few sticks, little actually penetrated. After the first violent gusts in the van of the storm the wind ceased with the lashing rain, and there was nothing to do but lie there and wait.

It was suffocatingly hot under the tarpaulin and there were plenty of mosquitoes; but we could reach the whisky bottle and some ready food. We swigged the life-giving fluid out of little mugs in a civilized manner, wriggled into our sleeping-bags, and, when we realized that we were not going to be absolutely soaked, began to enjoy ourselves thoroughly. We laughed and joked, supremely happy in our good companionship and in the knowledge that here we actually were, on the pelican island with two nights to spend, almost certain to have a feast of ornithological discovery on the morrow. The rain lashed down for an hour and a half, and then continued in a fine drizzle. We saw no reason to come out from under, despite the heat and the mosquitoes. Replete, tipsy, and warm in our sleeping-bags, we fell asleep.

During the night, when we woke, in the silence following the storm, we could hear a continuous deep hum from the pelican colony. The lake had stilled in a glassy calm, and the roar of waves was gone. The deep sonorous hum resembled no other sound I ever heard, but was evidently the conversational nocturnal murmur of thousands of pelicans. We guessed they were awake, and wondered if they could see, but decided to leave that bit of investigation till later, when it might not be so wet and uncomfortable.

At dawn the croaking of Fan-tailed Ravens and the cooing of innumerable Speckled Pigeons woke us. We crawled out into nice warm sunshine. Already, at first light, some pelicans were leaving, making for the feeding grounds at Lake Abiata 16 kilometres (10 miles) away. Cormorants in skeins also passed over, but would not fly direct over the pelican island; rather, a skein would split and fly round to reunite beyond. At 6.35 the first Black Kite arrived, from the mainland about ten kilometres away. It must have been on the wing soon after first light, and by seven o'clock it had been joined by two others, some Egyptian Vultures, and a Fish Eagle. These predators had all left the pelican island to roost, returning in the morning, while the Speckled Pigeons had roosted on the island to escape nocturnal predators on land.

We ate some breakfast and climbed to the ridge to see what was what. There was a very nice group of pelicans fairly close to the ridge-top, providentially with small young. We had sticks and some hessian to make a hide, but soon found that we could not push the sticks into solid very hard rock, where we had expected soft earth. So somehow we had to erect the ramshackle thing around us, and crawl forward till we judged that we were within easy range using it as a shelter, loaded with our gear, and picking up stones as we went to support the sticks. By good luck we reached a point only about thirty metres from the group of pelicans with very small young; and there we stopped, content.

Fortune had smiled on us, for our best-placed group all had very small young or eggs about to hatch. A few days either way and they would have deserted *en masse*, or moved off, their young shuffling after them. We did not know that then, but found out such details of behaviour later. When we had drawn breath, made ourselves as comfortable as we could, and sorted our gear into handy piles – using some of it to shore up the poles of the flimsy hide – we set out to make counts and note details of behaviour. No one had ever sat so close to a colony of Great White Pelicans before, and we were determined to seize our chance with both hands. In the hours that followed discovery flooded upon discovery, until we were sated with new knowledge.

We saw at once that the whole colony, of about 1,700 pairs, was made up of groups at different stages, some with eggs, some with small young, others with larger young, yet others displaying near the shore and evidently about to breed. Towards the top of the island a cohort of nearly fledged young showed that breeding had begun at least three months before. Other large feathered young swam round the island, making the communal fishing movements of adults, although there are no fish for them to catch in Lake Shala. Perching on high rocks, some of them flapped their wings in uncertain attempts to take off. In other parts of the island groups of black downy young, packed together in clumps that are for some reason or other known as 'pods', awaited the arrival of parents with food.

In such a situation there is a limit to the amount of new information that can be quickly grasped, so we concentrated on the group nearest us, with hatching eggs or very small young. Now we saw the ultimate marvel of the pelican's wonderful beak. A pelican's legs are situated very far back on the body, useful for swimming, but making it ungainly on land.

The eggs are incubated by placing the web of each broad foot over them, easy since all four toes are webbed into an expanse larger than my spread hand. When the egg hatches, it produces a tiny, helpless, pink, naked chick, half-blind and wobbling, that must somehow not be trodden upon and must be fed, or induced to feed, or it will soon die. The chicks are almost one and a half metres (four feet) from the pelican's eyes, if the bird is standing up. Yet, somehow, small quantities of nutritious food must be delivered to that helpless little thing, with a beak that seems hopelessly unwieldy for so delicate a task.

To do so, the pelican, with the head bowed, places its beak beneath its body, stretching almost horizontally back between the legs. The tip reaches the tiny chick or chicks, way back under her or his tail – for both sexes are equally adept at mothering small chicks. Some sort of fluid, perhaps digested fish, or even maybe a crop secretion, as in flamingos, flows down the channels on either side of the upper mandible, which is now beneath the lower. It collects in the nail-like structure, which when upside down is a tiny cup, at the tip of the beak. In the breeding season this nail-like structure is cherry-red; and the wobbling feeble little chick instinctively pecks at this cherry-red object, and somehow gets its first very necessary feed. The delicacy and gentleness with which a massive and apparently unwieldy instrument is used is marvellous to behold.

The pelican of legend was famed for her piety, because she was supposed to feed her young with drops of blood plucked from her own breast. One theory has it that these pelicans were not actually pelicans but flamingos, which do feed their young partly on their own blood, in a secretion. However, the old picture of the Pelican in Piety looks to me like a pelican, not a flamingo. These pelicans in Europe of old must have been either Great White or Dalmatian, more likely the commoner Great White. Great White Pelicans often or usually have a darker band across their breasts, caused actually by staining in the waters where they feed, but which could look like dried blood on the feathers. To an observer in, say, A.D. 1300, lacking any binoculars or other such aids, that cherry-red nail on the tip of the beak could look like a drop of fresh blood on which the young was feeding. Anyway, to us it seemed – and still seems – that this could be the origin of the fable. In any case it was another marvellous and highly specialized biological adaptation of what seems a thoroughly unwieldy and grotesque organ.

We sat on, making counts of clutch size, brood size where we could see them, and many other things. Watching the groups in display near the shore we saw that both males and females developed big swollen bright-coloured knobs of flesh on the forehead at the base of the beak. Ever afterwards we knew them as 'knobbers'. Since these knobs disappeared almost as soon as the eggs are laid, they were evidently pre-nuptial adornments used at mating time. Any knobbers seen in a flock of pelicans anywhere indicate that they are breeding not too far away. The American Great White Pelican at this time grows an extraordinary horn-like process on top of its beak; but I do not know what, if any, special adornments are assumed by the Dalmatian or Australian Pelicans.

Female and male Great Whites can be distinguished by the colour of their knobs; hers is bright orange, his bright pale yellow. They also have crests of different lengths, usually longer in the female, but that difference is not constant. The colourful and distinctive knob immediately enables males to recognize a female ready to mate. In both the American Great White and in our Great White such a female is assiduously courted by several males as soon as she alights in water or on land. One male seems soon to become dominant over rivals; and the pair then go ashore, and move along in what is called the strutting walk, ridiculous yet somehow dignified, head up, wings akimbo, stern waggling provocatively from side to side.

This ends with the female entering a group all at a similar stage. She settles immediately into the place where she will lay, and there the male mounts and mates with her. Thereafter she remains where she is, and he walks about collecting nesting material, which he either gathers from bushes, or steals from other pelicans – if they will let him. He collects a mass of it in his capacious pouch – yet another use for that extraordinary organ – walks back to his mate, and deposits it at her feet with a grave bow. She picks at it and fashions it into a crude, slight nest, in which she lays one to three, usually two, chalky-white eggs.

Thereafter she and her mate take turns in sitting on the eggs and tending the small young. Such turns may last from twenty-four to seventy-two hours, one partner being away feeding, while the other sits continuously. This continues even after the young hatch and must be fed at frequent intervals. If *you* eat a fish you digest it all in two or three hours. Not so a pelican. A bird which has been brooding small young for forty-eight hours can still regurgitate a nearly complete fish, move it about in the pouch, and re-swallow it. Normal digestion is in some way controlled or delayed; but I do not know how.

Bigger young become black and naked, and are hideously ugly. Then they grow a coat of black down, and collect together in pods of 10 to 30 or more. These pods move some distance from the place where they were hatched, leaving a vacant space in the colony where other pelicans can come and lay. Thus the same piece of ground is occupied through the season by two or even three different pairs. The total number breeding yearly at Lake Shala was always several times the number nesting at any one time. A colony of 1,500 pairs is made up of several groups, each of which started to breed more or less together. In a good year 10,000 to 12,500 pairs breed, with a maximum of 3,000 to 4,000 present at the peak breeding time.

Incredible as it may seem, in a pod of black downy young, all completely indistinguishable to human eyes and crammed together in an amorphous mass, each parent knows its own chick, and will feed that one and no other. Such young are left almost entirely alone; and one or other parent, returning to feed, alights in the open ground at the edge of the colony and searches about, moving from pod to pod, until its youngster is found. Then the parent often reaches far out over the mass of others, seizes the luckless chick by the neck, drags it forth, brutally shakes it as a terrier shakes a rat, and in this violent manner wakes it up and induces it to beg for food. The once comatose youngster then thrusts its head up into the parental pouch for nourishment. Incredulous, we saw this happen, and wondered how quite small young could survive such violent handling, not rarely, but every day.

Older young get their own back, and with interest. They can recognize the incoming parent far up in the sky. They run about, squalling and flapping their wings as if demented, often biting themselves, watching the parent as a hesitant full back judges the flight of a descending rugger ball. The parent alights, and at once runs away from its now frantic offspring, which pursues, apparently working itself into a state of dementia. Eventually he or she stops at a convenient place, and then the young one mercilessly seizes the parent's bill and forces it downwards. A big male youngster may be taller than, and half as heavy again, as his poor little mother. He forces her to a squatting posture, then thrusts his great beak first into her pouch and then far down into her gullet to get at the food. Once in, he opens his beak, which can be seen distending and stretching the parental throat as he struggles to obtain his food. The parent now cannot easily eject the young; and the young, with its head inside, cannot see what it is doing. When the food runs short the young one rises to its feet and drags the unfortunate parent about willy-nilly. Such feeds are only ended by a violent struggle, in which the parent strives to throw the young one out of its throat. How they avoid severe wounds is obscure. The best comment ever passed on this gruesome scene is that of Diana Powell-Cotton, who observed that those who find certain aspects of human motherhood disgusting had better try being a pelican!

All this we saw and broadly took in that first day; details sorted themselves out in later visits. By eleven, when we had been watching for some hours, we were for the first time and rather unexpectedly involved in what we call the midday rush. Pelicans cannot feed on Lake Shala, where there are no fish in the open lake, but must go thirty or more kilometres (about twenty miles) to other lakes to fish. Since they cannot rise to height without the aid of thermals, they can only take off after about nine o'clock; and from eleven o'clock to one they come in from wherever they have been gathering food in constant succession. It is one of the most spectacular sights in all nature.

We would first see an incoming bird high up in the sky, almost out of range of binoculars. We had no means of knowing where it had come from, or why it was quite so high; it often seemed unnecessary. It would hurtle downwards, first a little speck against the blue, gradually resolving into a great broad-bodied flying-boat bird, with partly furled wings, and lowered legs ending in huge yellow splayed feet, used individually to control direction like infinitely adjustable air brakes. Nearer the ground the downward speed slowed, and the bird circled the island several times in slow descending, infinitely graceful arabesques. The final circuit ended with a thumping of great wings to brake. Then our pelican was waddling about, supremely ungainly once more, to work its way into a sitting group to relieve a mate, search for its young in a pod, or be greeted by an importunate large young one bawling for food. This scene is at its best when most pairs have young; then the first large numbers begin to come in about ten o'clock, with a peak around midday, continuing till early afternoon. Then activity dies off till the next day, at the same time.

All this we saw that first day in our flimsy hide, exposed to the merciless sun, and barely able to move with cramp. Yet so absorbed were we that the time seemed to pass in a flash. We only felt like leaving when the activity died down, and most change-overs and feeds had taken place. Sated with a surfeit of new knowledge, and unable fully to comprehend everything we had seen, we withdrew to the only shade the island offered, a small acacia tree. Here we ate, and then went in the boat to look at two other steep islands nearby, later christened Sacred and Abdim's Islands, because of the colonies of Sacred Ibis and Abdim's Storks we found there.

At about four we returned to our flimsy hide, and watched the rest of the evening. Aerial activity had died down greatly, but almost all pelicans with young were feeding them. We could watch everything from the feeding of tiny naked youngsters to big waddling brutes, heavier than their parents, and could take it all in rather better than we did earlier. By 5.45 the colony was settling for the night, but just before we left a strong gust of wind blew, and a group of big youngsters made one of their first flights on it. They flew quite competently, circled, and did not attempt to land downwind. They knew how to fly almost as soon as they had to.

We withdrew to our camp. This evening, we were spared any threat of a storm, and we lay comfortably on the tarpaulin on our air-beds, in the open, as the stars came out. From the colony there came that deep sonorous conversational hum, punctuated by the high-pitched squealing notes of young. Later these died away, but that rich hum persisted. We ate, drank our whisky, and talked over the events of a staggering day. Neither of us will ever forget it, nor will the spectacle ever pall, no matter how often it is repeated. We were worn out with continuous mental effort and physical discomfort; but it had been so well worth it, a hundred times over.

Just before we slept, at 9.30, we went to see what was happening. The deep hum had died to a gentle murmur, and the voices of all young were stilled. It was a moonless night with stars; but one pelican flushed as we were creeping towards our hide and, when we shone a torch on the pelicans for a second, they were all awake, heads up, alarmed. They had not gone to sleep as might have been expected. Later observations show that they never really do go to sleep, at any hour of the night. There is always some hum, and if a human head appears incautiously over the top of a ridge there will soon be a panic. We crept out of the hide, and back to bed. We slept soundly under the vault of stars, infinitely content.

Next morning we did a little more observation, trying actually to count young in nests. However, it was immediately obvious that walking about in the open, even at reasonable range, caused panic, and we gave up after a few minutes. Statistical data were much less important than the success of the colony, which we alone had been privileged to watch at close range. We reached the shore again after a two-hour journey, and had to fight off aggressive Galla youths demanding ridiculous sums of money. I should have liked to have a good stock-whip handy; a *sjambok* would have been too short.

We came back again for another long week-end in the middle of my Nyala survey; and this time based ourselves at Lake Langanno Hotel for comfort's sake, going out in the boat for a day. That hotel, where I had first stayed with Bill Corcutt in 1963 *en route* to the Bale Mountains, became a haven of peace and rest for both of us in later years. It was one of a chain of tourist hotels, somewhat primitive, but adequate for the not too fastidious, started by Ato Bekele Molla, who did more for tourism in Ethiopia than any individual alive – merely by providing basic amenities to make himself a living. If ever a man deserved to make a fortune he did; no doubt it has all been confiscated now.

We spent one day on the island, and other days investigating other aspects of pelicans' lives. On our return from the first day we ran into a storm and had to diverge and hug the weather shore of the lake, under steep rocky slopes, the whole way, using up nearly all our petrol. Emil, who drove all the way, vowed that he would never risk it again; but we made it after $2\frac{3}{4}$ hours of wet anxiety, and feasted in the Lake Langanno Hotel that night. Later, the National Geographic Society gave Emil a magnificent red speedboat, and such alarming trips and nights on the island became unnecessary. Pelican-watching was more comfortable; but somehow it did not recapture the thrill of that first visit, all by ourselves on the little island, in the middle of the dark and stormy lake with our cockle-shell boat.

We afterwards visited the island many times, and gradually pieced together most details of the pelicans' lives. I came from Kenya when I could; and at other times Emil went, alone or with other friends, notably Ian Gibson. Emil eventually erected a prefabricated dexion-and-board hide, as all our flimsy hessian hides were trodden into the ground by the hordes of large young pelicans late in the season. That permanent hide is there yet as far as I know. Emperor Haile Selassie and Prince Philip have both been in it; and that incomparable Anglia Television cameraman, Dieter Plage, and I spent nights out there while he made a film, *Pelican Flyway*, which many of you who read may have seen. Dieter was known to us, by virtue of his gigantic appetite, as Dieter the Eater. Since his appetite did not stop at food, whenever we saw a pretty girl on the beach at Lake Langanno we used to urge him to 'get knobbing'. He never, so far as we could see, quite accomplished the strutting walk.

Familiarity with the pelicans never bred contempt. Each time I went there was something new to see and marvel at, while the glorious spectacle of the midday rush never palled. Some of what we learned has yet to be published. Emil counted the pelicans each year, and found that the total numbers breeding could vary from 3,000 to 12,500, with a maximum at any one time of about 5,000. What we know shows quite clearly that this little island is one of the most important colonies in the whole of Africa, for most of the other permanent colonies are smaller, and we do not know whether the huge colonies in the Rukwa swamps are regular, although there are plenty of biologists in the game department equipped with aircraft who could find out quite easily.

So far as we know, there are less than a dozen colonies of Great White Pelicans in the whole of Africa, the regular ones being on the Banc D'Arguin off Mauretania, Wase Rock in Nigeria, and the spectacular rock mountains of Abu Touggour and Kapsikis in Chad. Then Lake Shala, the Rukwa swamps, St. Lucia Bay in Zululand, and two islands off the South African coast. The largest colony is at Rukwa, estimated at 40,000 pairs; the others together do not total so many. Sporadic breeding has also occurred at lakes Natron, Mweru Ntipa in Zambia, and Ngami in Botswana. Recently colonies have been found in the lower Zambezi Valley, the Etosha Pan, and in Botswana.

We do not know how regularly these colonies are used, or how many pairs breed in some of them over a period of years. However, results from Lake Shala indicate that not every pelican breeds annually, so that the total population is greater than the sum of breeding pairs. Only Lake St. Lucia and Lake Shala have been watched for any length of time. The most astonishing colonies are those on the great isolated rocks of Kapsikis and Abu Touggour, where the pelicans breed on the summit of unclimbable conical blocks of rock, in one case 160 kilometres (99 miles) from the nearest water. Inaccessibility, it seems, is more important to them than proximity to a regular food supply. Inaccessibility also seems to control the timing of the breeding season – in the rains at Rukwa and in the dry season in some other colonies, including Shala.

In recent years, from 1968 onwards, Great White Pelicans have bred almost annually at Lake Elmenteita, Kenya. Up till then, the only East African breeding records were for the huge Rukwa colonies, never observed for any length of time, and the opportunistic effort to breed at Lake Natron in 1962, never repeated since. The Elmenteita colonies are the most easily accessible yet known, and as a result are endangered. It is all too easy for some well-meaning and enthusiastic bird-watcher who does not know how shy these great birds are to blunder out among them and put them all to flight. So far, fortunately, it has not happened often, though one visitor insisted, despite protests, in rowing out to the colonies in an open rubber dinghy; and this year several thousand pairs were caused to desert by another ill-judged visit by eminent people.

Here I have used my same old floating hide, designed for flamingos, to watch the Pelicans at point-blank range. At first I found them very nervous and if I had not known exactly what the signals of fright were, and given them time to settle as soon as one bird among thousands began to behave nervously, I could never have worked it up close. As it was, it took days, not the hour or so that was needed with flamingos. In time, and with care, I was able to beach my hide on the same island as breeding Pelicans; and once I was there, and they had accepted me, that was it! They paid no more attention thereafter, though one always must take care over the first approach, and if one pelican leaves its eggs it will not be back. Thus far I doubt if I have caused the loss of a dozen eggs among tens of thousands. A recording of a Bavarian brass band, which the flamingos would not notice, would not be appreciated by pelicans.

The chain of circumstances leading to the establishment of these colonies is so astonishing that it would be worth a whole chapter in itself.

The long-term ramifications of apparently unconnected events – beginning with the deluge of rain and the floods in the last three months of 1961 – would fill a book. I often quote them as the classic example of an ill-considered and well-meaning act of man, which at the time seemed reasonable, even beneficial, but whose end-results cannot be foreseen, even if they already begin to appear less desirable than at first seemed likely. I tear my hair at the knowledge that, when the trigger was pressed, as it were, there was no competent, dedicated team of scientists on the spot to observe the short- and long-term results. The few scientists available in East Africa were each working on their own project and without funds, research equipment, and time could not have achieved good results anyhow. Those of us who were there just had to piece it together as best we could, however inadequately. I said at the beginning that in the tropics one can often start with a completely clean page. Here was a nice clean page that has been covered with indistinct blots and unconnected scribbles, so that no one coming after will ever be able to understand it completely, however much money and research facilities are available.

In 1960–1 the little alkali-tolerant fish, *Tilapia grahami*, from Lake Magadi was introduced into Lake Nakuru. Although it does not eat mosquito larvae, someone apparently thought it would. The introduction coincided with the extraordinary rains of 1961, and the fish not only survived in the permanent springs, but spread throughout the lake. They have since attracted, and maintain, a huge population of fish-eating birds, cormorants, darters, herons, grebes, and spoonbills besides both Pink-backed and Great White Pelicans. The maximum number of Great White Pelicans recorded at Lake Nakuru is 35,000; and there are regularly 4,000 to 6,000 there. Annually the pelicans harvest about 2,500 tonnes of fish from Lake Nakuru, apparently without greatly diminishing the supply. The fish thrive, grow to a larger size on average than in Lake Magadi, and, if now seen by a naturalist who did not know where they came from, would probably be pronounced a separate species!

For years the pelicans fed on these fish without attempting to breed. Then, in March 1968, a big breeding colony of 4,500 pairs of Greater Flamingos developed at Lake Elmenteita. It failed completely because of attacks by only 17 Marabou Storks; but just before it failed about 120 pairs of Great White Pelicans came and laid on one island among the Flamingos. They too deserted when the Flamingos deserted, showing that they had been triggered to breed by the presence of the Flamingos. In July the Flamingos tried again, and this time no Storks appeared. Back came the Pelicans, and this time they increased to large numbers. Starting in the centre of the island, surrounded by Flamingos, they spread outwards until they took over the whole island. A fragile, long-legged flamingo weighing 2·5 to 3 kilogrammes (5 to 6 pounds) is no match for a behemoth of a pelican of 7 to 9 kilogrammes (15 to 20 pounds). If the pelican wants to lay an egg there, the flamingo just has to get up and go. I have watched it happen, often; it takes about twenty minutes, and one can sense the unfortunate flamingo's despair. The pelican means no harm; it is just much bigger.

Heavy losses among the flamingos occurred, but the pelicans built up to large enough numbers to create their own flock confidence. They never stopped breeding thereafter for two and a half years, until they suddenly deserted *en masse* in January 1971. As in that Natron colony in 1962, there were eggs, young at all stages, and displaying birds present. They just went, for no obvious reason but possibly because the lake levels were falling. There was no sudden die-off of fish at Lake Nakuru to account for it, and human interference was very unlikely, after two and a half years of continuous success. On 31 January 1971 I watched them with John Hopcraft, and they paid no attention at all when a hippo came out of the water and walked among them. Yet if we had tried it they would all have gone at once. Evidently they can be quite discriminating.

However, they again bred abundantly in 1973, 1974, and 1975, not in 1976, but twice in 1977. The first of the 1977 colonies, a huge one of about 8,000 pairs in two groups, deserted *en masse* in one night, apparently after a very heavy storm of rain which made the lake level rise rapidly. The pelicans breeding on low-lying islets may have been swamped and nervous. Those on higher islands lost confidence too, and went with them. Some came back in October, and bred in the same places; but most of these were unsuccessful.

From 1968 to 1975 if any unfortunate flamingos tried to breed on the islands where they were so successful in 1956–7 and again in 1966, the pelicans came too and blotted them out. There has been virtually no successful flamingo breeding here since 1968, though in 1977 a few hundred pairs successfully hatched young in October. Probably, however, they succeeded only because the pelicans that came to the same islands did not increase to great numbers. I recently watched both, from my floating hide, on an island where I photographed the flamingos in April 1957, twenty years earlier. There can be no doubt that the pelicans are attracted to breed on the same islands as the flamingos, for they choose these out of others available. They wipe out the flamingos by sheer weight, not malice. Evidently it is not unique, for Great White Pelicans have also been found breeding among flamingos, sometimes wiping them out, by Salim Ali in the Rann of Cutch and Hu Berry in the Etosha Pan. What goes wrong here in nature's grand design?

The Elmenteita and Nakuru pelicans have been studied in detail by Daphne Nightingale, especially their diurnal behaviour and fishing success. Her work has not yet been published. Detailed studies of behaviour in a breeding colony have been made by another observer, Simon Stagg, who lives in Kenya but has not discussed the pelicans with me; his work is also unpublished as yet. In his last years at Shala, Emil Urban ringed several hundred young pelicans, and attached coloured streamers of plastic to their legs. They have been recovered or sighted in the swamps of the Sudan, northern Ethiopia, and in Kenya, showing that they disperse rapidly and widely from the island at Lake Shala as soon as they can fly, and do not just go to the favoured feeding ground of the adults at Lake Abiata.

So the Great White Pelican, almost an unknown species in 1965 when we first found the Shala colony, may now be the best known of all the pelicans possibly excepting the American Great White Pelican. But it all really began about a decade ago, when Emil Urban and I, replete with food and whisky, lay roaring with laughter under that inadequate tarpaulin on Pelican Island, while outside the Almighty did his worst on the wings of his storm.

P. TISCHLER

12. Tigers

I saw my first wild Tiger at the age of twelve. I was then at school at Ootacamund, in the Nilgiri Hills of South India. In those days (and still to some extent now) Ooty was surrounded by beautiful rolling country of short grass, with patches of forest, called *sholas*, in valleys and depressions. In those days we did not have to go far from Ooty to find wild country, with Sambar and pig in the *sholas*, leopards and Tigers, langur monkeys and the beautiful chestnut-and-black giant Malabar Squirrel, an animal as big as a cat, which has remained one of my favourites ever since.

Occasionally, when I had saved enough pocket money, I used to hire a bicycle in the Ooty bazaar, unknown to my parents, and go out as far as possible on the downs with a friend – even I had the sense not to go quite alone on these jaunts. Thus it happened one day that I went to a place called Governor's Shola (because, I believe, Tigers were baited into it so that visiting governors could easily shoot them) with an older boy called Harrison. We both possessed quite powerful airguns, and our objective was to bag a Grey Jungle Fowl, a thing neither of us ever succeeded in doing, though their crowing calls were loud in any *shola* at dawn, and they could be induced to approach within range by a skilful imitation of the clarion voice on blades of grass held taut between the top joint and ball of each thumb.

On the Ooty downs there lived a tribe of semi-nomadic people called Todas; I believe they live there still. They kept large herds of buffalos, which could be aggressive and which we often had to avoid by getting under river banks too steep for the threatening bulls to negotiate, before their own little-boy herdsman came and drove them back. In those days these buffalos were often kept at night in stone-built pens, with walls some one and a half metres (five feet) high and a metre (three feet) thick. The manure that accumulated in these pens over years was periodically cleaned out and thrown over the wall; and where it fell there grew a dense shrubbery of blue-flowered, bitter-leaved plant which I now know to be a species of *Vernonia*. We called it goatweed, and it produced a dense cover about a metre high.

One such Toda buffalo-pen stood on the very edge of Governor's Shola, where I had been bird's-nesting several times. The manure was thrown out towards the *shola*, and as a result there was a patch of luxuriant goatweed about twelve and a half metres (fifteen yards) square. We knew from the signs of scratching in the rich soil among the weed that jungle fowl came out of the *shola* to search for grubs. So on this day we crept down to the buffalo-pen and very quietly made our way up to the wall. It was about four in the afternoon, and we intended to try for a jungle fowl and then go home before anyone started worrying about us.

We reached the wall undetected and peered cautiously over. There was nothing in sight. However, in the goatweed we heard a slight rustling which might have been made by a jungle fowl. So I took a stone from the top of the wall and carefully lobbed it into the weed beyond the sound, so as to make the bird run towards us, while Harrison levelled his air rifle to take what we knew must be a quick shot.

But it was not a jungle fowl. To our unutterable surprise and, let us admit it, abject terror, a great tawny head reared out of the goatweed. Doubtless the Tiger, for such it undoubtedly was, had a curious and in no way malign expression on his surprised face. But we did not wait. I do not know what the under-fourteens record is for a 200-metre up-hill dash, but I bet I broke it *ek dum* (at once). I found myself lying panting on a grass slope near the top of a down, looking back, relieved to see no great ravening beast in my footsteps. We did not go back to the buffalo-pen, for we reasoned, to calm our nerves, that there would not be any jungle fowl now anyway. Mounting our bicycles we rode soberly home, and never said a word to our parents, lest more severe and effective prohibitions of such exploits be thereafter enforced.

At that time, and for many years afterwards, my sole aim was to return to India and become a tea-planter. I knew that such a job meant living in a pleasant climate, and plenty of chances for shikar – big-game shooting – which was not then frowned upon at all. In fact, it turned out otherwise, for after I had taken a pass degree in Biology at St. Andrews University I could not get a job with the tea-planting firms I had in mind owing to the depression, so I went on to do Honours, without much enthusiasm at the time, but with no regrets at all now. The rather indifferent second-class Honours degree I obtained enabled me to apply for a job in the Colonial Agricultural Service as well as a tea-planter; and the Colonial Scholarship was offered to me ten days before two tea companies offered me jobs. I suppose possible employers had formed a better idea of me than I had of myself. At the time I was doubtful about it, but my father, who had more sense than most people I have met, advised me that in hard times I should take what was offered rather than hope for what might never materialize. There were at least ten others ravening for the job if I did not take it; so I did, knowing that it meant good-bye to the India I longed to return to, and to Tigers – or so I thought. However, I consoled myself with the feeling that it might take me to East Africa, and in the end it did. My life has been greatly enriched by that fall of the hand of fate; and I *did* see Tigers again after all.

My mother died in 1953, and after spending some time with my brother and me in East Africa, my father returned to live in India, the only place where he had ever been really happy in his long service with the Indian Civil Service. We were relieved, because he could rent a comfortable bungalow in his beloved Nilgiris from a friend, Colin Primrose, for the princely sum of forty rupees (about £2) a month. He could have a car and travel about as he wished, and could get an excellent servant, Shanmugham, to attend to his quite simple needs. It also meant that we could go to India to see him; and in 1958 I went.

That was a full year for me. I visited the United States for the first time to learn about Range Management, got married for the second time, began building the house I have lived in ever since, and spent six weeks in India with my father and, for some of the time, with my brother too. The bungalow in Kotagiri, where I first went to school at a convent (having walked out of an American Mission School when they told me 'Study your Bible, little boy'), was simple but comfortable, and there was a good chance of seeing and shooting a Tiger. I had wanted to shoot a Tiger since I was a boy, and even in 1958, when there was probably only one Tiger for five or ten that existed in 1929, when I saw my first. Up to 3,000 Tigers a year were said to be killed on shikar by visiting sportsmen and residents; and it did not seem such a terrible crime that I should shoot one too, as it would be regarded today. So I had the dual objective of bagging a Tiger and seeing more of the India I recalled from boyhood.

Two things stand out in my mind. At the age of eight I lived in a bungalow on a stock farm at Hosur near Bangalore; and on going back there thirty-five years later I walked straight into the bedroom that had been mine, and had the strangest feeling that I had come home, for the first time in my life. Second, I remembered areas of the Nilgiris along the Dodabetta road between Kotagiri and Ooty where as a boy I had looked for birds' nests with boon companions in quite untouched forests. All gone now, hacked down, cultivated to potatoes until the soil was eroded to uselessness, and now reverted to dense scrub of hill guavas, on whose succulent small fruits we used to gorge – and I did gorge again. Governor's Shola looked much the same; but the downs between it and the Ooty were now cultivated, the Todas largely gone; and no Tigers had been heard of there for some years. There were not any governors left to shoot them anyway.

My first essay at bagging my Tiger was in the Moyar Valley at the foot of the now disused Gudalur Ghat road. Colonel Pythian-Adams, who was just retiring as President of the Nilgiri Game Association, had a little bungalow at Dilkush which he lent to me; and for a week I was there alone, and hunted Tigers assiduously. There were at least three within 16 kilometres (10 miles) of the bungalow; and on the second evening one killed a cow about one and a half kilometres from it. I hurried to the spot and sat on an excruciatingly painful branch until near midnight, then gave it up and went back to bed. The Tiger never took even one mouthful from the kill.

Thereafter my routine varied little. I hired the statutory bait buffalos for, I think, four rupees a day, and they were tied out by a *shikari* (hunter) in places where Tigers were likely to take them. I sat up nightly over one or the other, having parked my hired car a kilometre or so away. No Tiger ever came; but one such evening I had a rich reward of other things. After I took up my seat, about five in the evening, a troop of langur monkeys went to roost in the thick trees about my perch, and a group of the lovely Malabar Squirrels fed on the bark of the trees until dusk. The langurs coughed, scratched, and conversed with low calls as they settled to sleep, and then suddenly set up a loud outcry of alarm. 'By God, this is it!' I thought; 'A Tiger is coming.' It was not a Tiger, but a little Leopardess, who sprang up the bank of the river and walked daintily along a game path, tail curved over her back, head up, jaunty as you please, until she was lost to view thirty metres beyond. The buffalo, chewing the cud contentedly, never stirred; and she did not see it. There were no further alarms or excursions until, at one o'clock, I climbed down from my tree, waking up the sleepy langurs, and made my way through the silent moonlit jungle back to my car and bed.

The buffalo was still alive, chewing the cud, in the morning; so was his companion, tied up a kilometre or two away. Both had by now earned two-thirds of their purchase price in hiring fees; and it occurred to me, perhaps – but I think not – unjustly that the shikari knew the spots that were likely to be avoided by hungry Tigers, and was not only taking his pay from me but getting a cut from the buffalo owners as well. I am a suspicious brute in such matters. Anyway, I decided that I had better see what I could do on my own.

It had been my custom to sleep until nine, eat a huge brunch of fruit, that delectable dish Country Captain (chicken in a curry sauce with chips), and plenty of coffee, and then set off into the jungle to watch birds and beasts until about four when I had a meal before the nightly sit-up. Leisurely and silent progress through the jungle, which was actually quite open savanna forest with fire-resistant trees and spear grass, was both easy and pleasurable for one who was by now a very experienced hunter and tracker after many years in the far richer game country of

East Africa. The jungle was full of deer, Sambar, Chital (surely a candidate for the title of the world's most beautiful ungulate), and at least one herd of the majestic Gaur or Indian Bison, certainly the world's most spectacular and impressive bovine. It was also rich in birds; and I have ever since wanted to spend time watching the varied species that came to feed on the nectar of the beautiful orange flowers of the *Dhak* tree, *Butea monosperma*. There were at least three types of parakeets, glorious golden-green leafbulbuls, mynas, and others, and a little striped squirrel or chipmunk which took the flowers in its tiny hands like a ceremonial goblet, tipped them up and drank the nectar with an expression so comically ecstatic that I at once thought of Coleridge's line 'For he on honeydew hath fed, and drunk the milk of Paradise'.

In the afternoon these jaunts usually used to lead me to a shady track demarcating a forest reserve at the very foot of the Nilgiri slope. It passed right through a dense patch of Lantana bushes, and, having read Jim Corbett's stories, I studied it for signs of Tigers. I soon grasped that at least one Tiger regularly passed that way, probably every four to six days, making the scratch-marks that Tigers do make, defecating at intervals, and generally giving himself away. He – it was a big male by his old tracks – had not passed for three or four days, and I thought that in my last three days in camp he ought to pass. So, thereafter, I went daily to the fire-track to read the signs; it was a nice afternoon walk anyway.

On the second last day of my stay I found the 24-hour-old tracks of a large male Tiger entering the Lantanas on the fire-track at its southern end; I had not been there the day before but had explored the northern end and other slopes near. I thought at once that he would be taking cover in the heat of the day among the Lantana bushes, which were about six metres (twenty feet) tall, a contiguous thicket of thorny shrubs coming to within a metre or so of the ground and carpeted beneath with tinder-dry leaves, hopeless cover to work through silently. However, two hundred metres into the thicket a dry watercourse shaded by bamboos came down from the hills above, and this seemed the natural, obvious way for the Tiger to take to reach his shady retreat. So, stepping from stone to stone, and searching any sandy patches and game tracks, I went slowly up the watercourse.

I found nothing; and in fact I do not think the Tiger had gone that way at all. The watercourse narrowed and petered out at the base of the slope; and seeking somewhere cool where I could rest and watch I climbed about thirty metres (a hundred feet) and reached a big rock, from which I could look down on the ground below me.

The fire-track was roughly the base of a triangular bay in the mountain slope, all densely covered by Lantana. Right in the centre of the thicket there was a clump of forest trees, and on the top of these trees a troupe of langur monkeys were looking down at something beneath them, barking and jumping about from branch to branch. I had often seen or heard African baboons and monkeys behaving in a similar way in fear of a Leopard, and I thought to myself 'Aha!'

I did not know the ground well, so I again descended the watercourse till I judged I was level with the trees and then crawled two hundred metres along indistinct game trails towards the langurs which I could still hear barking. I was soon lacerated and bleeding from multiple scratches, and, try as I would, silent progress was impossible. When I approached the trees the monkeys barked again and left; and then I heard the heavy tread of a big animal moving slowly off, not bothering to conceal its movements. I had been in a similar close proximity to a Leopard only a year or so before, and I knew that this was no Sambar or pig, which would have bolted suddenly and with a crash. It could only be the Tiger; and a *frisson* went through me when I realized he had lain till I was within thirty metres or so before moving off.

Shortly I found his bed, on a cool soft patch of leaves under a shady tree; it was still warm from his body, and a tawny hair or two proclaimed his recent presence. He had slept here, in another bed close by, twenty-four hours ago too, and probably did it often. I realized that pursuit through the Lantana now would be useless, rose, and found a much easier way out into the fire-track, by which he must have entered, though it was grass-covered and he left no tracks.

'Now', I thought, 'I've got you, my boy!' I had only to tie up my two buffalos, one at either end of the fire-track, and for sure he would kill one or the other. However, I was still not quite sure that the Tiger was in there yet, or whether he had left on being disturbed. To ascertain whether he had entered that morning I went along the fire-track towards its northern end, where it debouched upon a stony hillside with a rock outcrop that I had already thought of as a good ground machan or place to sit. As I was approaching the opening I suddenly heard bulbuls chattering vigorously in the Lantana about twenty metres to my right. They could have been chattering at the Tiger, or a mongoose, or a snake; but I lay down on the track and searched intently below the thick shrubs. I could see nothing, though I half-expected to see a red leg, or even to find myself staring into the Tiger's eyes at close range. After a few minutes the bulbuls stopped chattering, and I stood up and moved quietly on.

I crossed a small patch of sand just outside the Lantana, but there were no Tiger's tracks upon it. So I went on two hundred metres till I came to a bigger sandy area on the track. There, freshly imprinted, were that morning's tracks of a big male Tiger coming in from the north. So it had been he whom I had disturbed from his bed after all; and he was probably still in the thicket. I hurried back, meaning to start making a hide at the rock outcrop. I reached the first sandy part of the track; and there were his tracks on top of mine, only a few seconds old.

It had indeed been he at whom the bulbuls had chattered. I had twice been within twenty to thirty metres of him, unable to see him. He had watched me off the premises, and then made good his escape. I brought up my buffalos that evening, and sat all night, sleeping at intervals, in the rock outcrop hide; but he had been warned, and at once put a safe distance between us, so that after a final futile day's searching for his traces I gave him up. I did not mind. I felt I had had a good try, learned something about Tigers, and been beaten by a very clever adversary.

I did in fact shoot my Tiger on that trip, later, in the Biligiri–Rangan hills of Mysore. It was a good hunt, and it strengthened my respect for the Tiger, who ate, in a week, half a man, part of a bullock, the back leg of a buffalo, and part of a cow. I made it 159 kilogrammes (350 pounds), some of it eaten by a Tigress who was with him for two days and whom I actually thought was the man-eater. He never returned to any of his kills; and after sitting up for five consecutive nights I shot him in a beat. He was 2·9 metres (9 feet 3 inches) long, huge, immensely fat and powerful. No zoo Tiger ever looked like he did as he strode slowly out in the sunlight in long grass only two hundred metres ahead of the beaters, to fall dead with one merciful shot. If anyone decries me for doing it, they should bear in mind that at that time there were still many Tigers in India and that the death of a man (over whose putrefying remains I sat for two nights) made it certain that if I did not hunt him down someone else would have done so. I built my house in Karen round his skin; but in 1971 it was torn off the wall by burglars, who cut off his head (secured to a steel bracket) with my wife's kitchen scissors. The insurance company paid me £300, as the value of a mounted Tiger skin at that date, when the crazy fashion for cat furs worn by women had endangered some of the world's most magnificent beasts. The burglars could not, however, take away the memories of another good and this time successful hunt, in which I saw and shot the second wild Tiger I had seen in my life.

In 1960 I revisited India, for the last time. My father had lived there happily, visiting Kenya at intervals to see us, for eight years, when, still in full possession of all his faculties, he was stricken by cancer of the liver. This particular form of cancer kills slowly, almost painlessly, but relentlessly, and, at the end, very distressingly. At any rate, I went back there to spend some time with him in his favourite haunts before the disease so weakened him that we could not enjoy the sort of life we both enjoyed, which was going about from place to place in the bush, climbing mountains and watching wild things.

We chose at first to spend a week at a forest bungalow at Guddesal, in South Coimbatore. I had no wish to shoot any more Tigers, which in the short period since 1958 had become more sharply threatened by poisoning with organochlorine insecticides – which Indians had learned to use for this purpose. However, I took out a licence, and spent some time hunting in rather a desultory manner for Sloth Bears, common enough even then; I had seen half a dozen when hunting in the Nilgiris and Biligiris two years before, but I did not care whether I got one or not.

My father's greatest joy was climbing mountains. He was then seventy-one, had the fell disease eating into his remaining strength, but was still game to climb the rocky peaks that stood about Guddesal. After a day's rest in the bungalow we left, early in the morning, to climb one just behind the bungalow. It involved about six hundred metres (two thousand feet) of climbing, and we took with us a local *Sholaga* (forest-dweller) to carry breakfast and be our guide. We could barely communicate with him, but we got along well enough.

Sitting on the top, in the clear air of a glorious cold-weather morning, I saw a pair of Peregrine Falcons playing about the 90-metre (300-foot) shelving precipice below our feet. My father was naturally tired and wanted to stay on top so I, who had never seen the nest of an Indian Peregrine Falcon, left him there and climbed down the cliff face, from ledge to ledge, searching in my practised way for the tell-tale mutes that would mark the nest ledge. I did not find it; and at about 10.30 rested on a broad grassy ledge before re-ascending to the top.

Some 150 metres (500 feet) below me was a valley running north, covered with savanna woodland of now naked white-barked *Anogeissus* trees and tall *Cymbopogon* grass. Down in that valley the grass would have been over my head, but from my perch I could scan it easily for any wild animals it might contain. A cart track ran along the further side of the valley, and came into the field of my searching binoculars at intervals. There was nothing on it, as was to be expected at that time of day. And then at a bend, in a patch of sunlight, there was suddenly something red. A monstrous, magnificent, full-grown male Tiger, bulging with his last night's meal of buffalo or whatever, making his slow way towards his daytime resting place. As he crossed a patch of sunlight, in the full blaze of the morning, he epitomized the allegorical words of Blake, 'Tyger Tyger, burning bright'; the most magnificent animal I ever saw. But he was not in the forests of the night, but in the full fire of the morning sun.

He was full fed, plodding slowly along, but never for a second did he relax his caution. Approaching a bend on the track that he could not see round, he slowed to a careful crawl, advancing almost centimetre by centimetre with his ears laid flat back, till he could safely proceed. Satisfied, his ears pricked forward again, and he plodded heavily on. His tread was almost lumbering, his full belly swung from side to side; but he was majestic, and I had him all to myself.

For twenty minutes I kept him in view, scarcely lowering the binoculars from my eyes. Every few minutes he backed into a bush, raised his tail in a graceful arch over his back, and left a little squirt of urine at eye level. Then he would demonstrate the purpose of this scent-marking by himself sniffing up and down, as I have often seen my dog sniff, at another bush. Twice he left the track a little way, crouched, defecated a little, and scratched with his hind feet, making the tell-tale scratch-marks that reveal where and how often a Tiger walks and are his one great weakness to those who would hunt him fairly on foot. He worked his way thus along the cart track for about a kilometre and then went into a dense patch of forest, about a hectare (two and a half acres) in extent, at the head of another valley, leading down to the Guddesal bungalow. There I supposed he would stay for the rest of the day, for there was no other obvious place in sight and I did not see him reappear.

In those twenty minutes I learned more about Tigers than in all the rest of my life. Admittedly, I never have had the chance to study them extensively, but it is not given to everyone who tries to watch a full-grown male Tiger in broad daylight, on an open cart track, for twenty minutes. The hunter sits over a Tiger's kill, and if lucky sees his quarry for a few moments over rifle sights before he ends the encounter, either with a bullet or an incautious movement. Had I been on the track myself I should at best have seen the Tiger face to face for a second or two. More likely he would have heard me coming and withdrawn into the long grass before I ever saw him at all. But from my elevated perch, where he never suspected my presence, I had a perfect view of him and all he so revealingly did, in glorious colour. I knew I would never forget it; and every detail stands out in my mind's eye today.

I went back to the hilltop and told my father what I had seen. It had not occurred to me that he had never seen a Tiger. He was acutely disappointed for, although he had spent many years in India when he was younger, and had spent time on shikar in the days when Tigers were ten times as common, he had never actually seen one. Once he had been close enough to a tiger to hear it breathing, but could not see it. I knew how that could happen from my earlier experiences in the Nilgiris. However, I felt that we were not done yet. The Tiger would assuredly go to rest in that *shola*; and since it was on our way back to Guddesal I hoped that I might be able to place my father strategically, and gently move the

Tiger out towards him. When he heard me on the cart track and in the *shola* his natural line of retreat would be down that little valley.

We dismissed our guide to his home, and he was heartily glad to go, for it was now hot. The haversack was empty of our breakfast and no chore to carry anyway. We descended to the track, where I walked slowly along, visiting as many as possible of the places where I had seen the Tiger do something, noticing what I could. His fresh scratch-marks where he had defecated were specially interesting, for they showed exactly what an hour-old scratch would look like. I could not detect any of the little squirts of urine, for of course they had dried, and I cannot smell like he could, but I thought that in some places the smell of cat was unusually strong.

Two hundred metres from the *shola* we left the track and sought a viewpoint for my father. We found a good rock where he had a fair chance of seeing the Tiger crossing a large open space of shallow soil, where the grass was relatively short. I then returned to the track and began to carry out my plan. We had done all this silently, and downwind of the *shola* (though Tigers, like all the big cats, have a poor long-distance sense of smell anyhow), so that he had no reason to suspect at all. I had high hopes that I would show my father his first, and certainly, alas, his last, wild Tiger.

But the great beast had yet another lesson to teach me that day. I came back along the cart track treading rapidly, and coughing once or twice. I envisaged his ears pricking as he lay in his bed, and thought he might cautiously retreat as he heard me coming. Level with the *shola*, I broke my way noisily through a screen of long grass and shrubs into it and proceeded to snap off dead branches as if I were collecting firewood. No sign, for some time, as I moved about, and then bulbuls began to chatter again. The Tiger was there, within fifty metres of me; and he was on the move.

The shrubbery was dense inside the *shola*, and I could not see him at all. We moved slowly about, like two boxers in a big ring, I trying to keep him moving in the direction of my father, he simply avoiding me with consummate skill. I never heard him once; he was moving cautiously, silently, not at all like that other heavy-treading Tiger in the Nilgiris two years before. His position was betrayed by the calls of birds, but I could not get him to leave. He knew, doubtless from long experience, that he was being driven; and he was damned if he would go.

Round and round we went for a quarter of an hour, perhaps more, I always trying to keep him shifting towards the valley. Then at length the bird-calls that betrayed his whereabouts stilled, and I guessed he had finally gone. I searched a little more, but in the mass of dry leaves found no tracks; for that matter I had found no single pug mark on the cart track along which I knew he had walked for twenty minutes – he had stuck to the grass in the centre or on the verge. So I emerged again and sought my father, still hoping I might have been successful in moving the Tiger the right way. But he had seen nothing.

I have never seen another wild Tiger and do not suppose I ever shall. Since that day they have been further decimated until a huge project, one of the biggest ever mounted by the World Wildlife Fund, has been set up to try to conserve the remnant population, estimated at less than two thousand, fewer than those reported shot *annually* in India in 1958. If the Tiger survives it will be a miracle, for the population of India is still increasing by leaps and bounds, and must go on doing so for many years despite Mrs. Gandhi's draconian sterilization decrees. Tigers survive as a precarious remnant population in a few National Parks and game reserves, and elsewhere in Burma, Malaya, Java, and even perhaps near the Caspian Sea. They have been studied in detail in the Kanha National Park by George Schaller, who has written a fascinating if rather dry book, *The Deer and the Tiger*. I have read many other books about Tigers, the most graphic being those of the late Jim Corbett, who wrote about the man-eaters he hunted in Kumaon in the twenties and thirties when I was a boy. Even in these I have never read of anyone who saw a wild male Tiger in his prime, in daylight, for longer or better than I did that day near Guddesal.

D. TISCHLER

13. The Whale Shark

The Whale Shark is the world's biggest fish. I suppose that is why it is called the Whale Shark, as otherwise it does not greatly resemble a whale. Whales are, of course, warm-blooded mammals, not fish at all. The basking sharks that appear in summer off British shores are big; but they are just tiddlers compared to Whale Sharks. A big Whale Shark may be 15 metres (50 feet) long, and weigh many tonnes, far more, for instance than an elephant, though not as much as a big whale. Blue Whales are the largest animals ever known to have lived.

Rather little seems to be known about Whale Sharks. They are solitary creatures that wander widely in warm seas. Occasionally, according to the books, a seagoing steamer strikes one and no doubt kills it, leaving no trace. I never even knew of their existence until I read about one that followed Thor Heyerdahl in the *Kon-Tiki*, right in the middle of the Pacific. They seem to come to such floating objects moving slowly, since one also gave the intrepid Dr. Bombard, who travelled across the Atlantic alone in a rubber dinghy, some moments of anxiety by rubbing its back on the bottom of his frail craft. All those who have seen them at close range seem to agree that they mean no great harm. They are not man-eaters. They are just big, very big indeed.

Whale Sharks are not too uncommon off the Kenya coast. However, they are usually seen briefly by charter fishing boats some distance out to sea. The continental shelf is very narrow off the coast of East Africa, and 16 kilometres (10 miles) out to sea one may be in water hundreds of metres deep. The Whale Sharks encountered by these big-game fishermen are often referred to as Basking Sharks, which they are not. They live near the surface, but this is not because they want to bask in the sun. They are simply there to feed on small surface-dwelling organisms, and the fact that their fins and even parts of their bodies may emerge from the water from time to time is nothing to do with sun-loving. Fishermen intent on trolling for marlin or other large game fish see the Whale Shark only briefly, as it submerges to avoid the boat, or swims past some distance away.

It is a strange thing, but all the very biggest creatures of the sea live on small organisms. The largest whales are the big baleen whales, the Blue, Fin, Right, Sulphur-bottomed, and others. They are all monstrous creatures, the largest some 37 metres (120 feet – 40 *yards*!) long. All these huge whales live on masses of tiny crustacea collectively called krill, which abound in some cold seas and can be harvested by a filtering mechanism of plates in the mouth. The curve of a baleen whale's lips resembles that of a flamingo's mandibles. Evidently this kind of curve can evolve in totally different creatures which just happen to feed in the same way – by filtering small organisms from the water. The flamingo actively pumps water in and out, but as far as I know the whale just swims along. The only very large whale which is carnivorous is the Sperm Whale; and the biggest males of these are nowhere as big as the larger krill-feeding species. Hermann Melville, in his book *Moby Dick*, exaggerates.

Sharks belong to the order of cartilaginous fishes without any real hard bones. They are generally dreaded by human beings, not wholly without reason, though modern scuba-divers have shown that they can usually be driven away quite easily. Here again, all the biggest sharks are not carnivorous in the sense that they kill large prey. The Basking Sharks one may see off the west coast of Scotland feed upon plankton. They appear off the coast of Ireland in March, migrating north from the Mediterranean and North African waters, and have reached the north of Scotland by June. They swim along slowly, at about two knots, with their mouths constantly open, sieving plankton from the water through their multiple gill slits. They may be up to 12 metres (40 feet) long, and are very heavy and powerful. However, one could swim close to them without the slightest risk of being attacked.

Likewise, among the rays, the largest is the Manta Ray, sometimes called the Devil Fish. It too lives on small organisms. It is a most extraordinary-looking creature, and its body, broad, flat, and nearly square, may be up to 12 metres across. Imagine a skate about the size of a squash court and you have it. In front the mouth, which is large and oval, like the radiator grille of some modern motor cars, is contained by two long, forward-projecting horns. Perhaps these are used to guide food into the mouth as the fish swims slowly along. I have only seen small parts of Manta Rays off the coast of East Africa, usually just a little of the edge of the 'wings', the outstretched, gently undulating lateral parts of the body with which the fish swims, like a skate. I hope some day to get a better view. Manta Rays too are not dangerous; they carry no lethal sting in the tail, as do some of the other big rays one may see. If you harpoon one, and in its desperation and fear it jumps out of the water and lands on your boat, squashing you like a fly with its sheer weight and size, you have only yourself to blame. You could have left it alone.

* * *

We spend about two months of the year at the Kenya coast. The house we occupy, and which some day we may be allowed to own, stands close to the coral point where Mida Creek drains into the sea. In front is a broad lagoon in which there are many coral heads and different types of reefs. One could spend a lifetime there and still not understand all of it. We found the house derelict in 1966 and restored it to a habitable condition. We call it *Zawadi*, a Swahili word meaning prize or reward. A Fish Eagle comes and catches garfish from the coral point. Sitting there in an evening I have once caught a brief glimpse of a Dugong, a marine or estuarine mammal now persecuted almost to extinction and unable to cope with modern nets. Being a mammal, if it swims into a nylon net which it cannot see it is entangled and drowns.

In Mida Creek we keep our small boat, called *Aonyx*, the scientific name of the African Clawless Otter, which you met in Chapter 10. Since I am a keen fisherman, and since I have no claws myself, it seemed appropriate; though I must admit I cannot eat crabs or mussels – they disagree with me. We have had two such boats, the first, *Aonyx I*, a Dell Quay Dory 4 metres (13 feet) long, later replaced by *Aonyx II*, a Boston Whaler of the same length, and a very much better boat. In these little boats, when the sea is calm, my son Charles and I go well out to sea to catch fish. Most of our catch is small tunny and bonito, which are not very good to eat, but we sometimes also catch kingfish and barracuda, which are. We caught two sailfish in 1970, one of them on a handline. However, a sailfish is such a magnificent, highly specialized marine organism that I am never very sorry if one gets off the hook. Charles thinks differently.

Charles used to have his school holidays in August and at Christmas. In August the south-west monsoon is blowing, and we can seldom go out beyond the reef. If we do go, we have our hearts in our mouths as we return, as we then often must ride the back of a swelling breaker, engine going strong, on the 'let 'er ride' principle. I am sure a good seaman would tell us it is very dangerous, but I do not pretend to be a good seaman. In December it is quite different. Then one can have mornings of almost flat, windless calm, when the reef presents no hazards whatever, especially at high tide. You could hardly kill yourself if you really tried.

On 31 December 1972 we had gone out in *Aonyx I*. An engineer friend of mine, from whom we later bought *Aonyx II*, described her at that time as 'doubtfully seaworthy', but she was all we had, and she floated. It was a morning of flat calm, with a very gentle breeze barely rippling the sea. We went out through the *mlango* or 'door' in the reef close to Whale Island, which is about one and a half kilometres off shore, and then on to a spot about five kilometres out to sea where upwelling water often results in a concentration of fish. On the way out we had an unexpected bonus in the shape of a Wahoo, *Acanthocybium solandri*, weighing 5·5 kilogrammes (12 pounds), said to be the fastest fish in the sea, and good eating. We reached our fishing ground and cruised about for an hour or so. We caught nothing; but we saw a rare bird, a Persian Gulf Shearwater, one of very few ever recorded in Kenya waters. If we do not catch fish in an hour or so we do not stay out, but go back home to breakfast. So, on this day, we were trolling back towards Whale Island at about nine in the morning. What little breeze there had been had now died, and the sea was glass-calm and very clear.

Not more than a kilometre or so outside Whale Island we saw a monstrous fin breaking the surface of the water. It was clearly a tail fin, for about four metres (thirteen feet) in front of it a smaller dorsal fin was also breaking the surface. We could not guess what it was, but approached on a parallel course. When we came to within thirty metres we saw the fin belonged to a Whale Shark. We could clearly see the vast grey body, dotted with white spots, sinking slowly into the depths as we approached. We judged it to be at least 7·5, possibly 9, metres (25 or 30 feet) long. Not a very big one as Whale Sharks go, for they can reach 15 metres (50 feet) in length, but at least twice the length of our boat, and weighing many tonnes.

Naturally, we thought the fish had disappeared for good, and congratulated ourselves on even getting a glimpse of it so close inshore. There cannot have been more than a hundred metres (fifty fathoms) of water under it, perhaps less. We went on fishing, and caught a small striped bonito, guided to it by the circling of a single Lesser Crested Tern. Then, when we were just under a kilometre off Whale Island, the great fin broke surface again. The fish appeared to be cruising slowly towards us. Accordingly, I positioned the dory approximately in its line of advance and – turned off the engine.

We were now floating silently on a glass-calm sea with the huge fish coming straight towards us. We stood up, and I took a good hold on the starting cord of the engine in case we had to make a bolt for it. I guessed that the fish had been frightened by the noise of the outboard propeller earlier and had submerged briefly, to rise and continue feeding later. With no engine noise, it was clearly unperturbed and came on, almost straight towards us.

It passed only about ten metres away, its huge bulk clearly visible in the clear water. As it was moving so slowly we could make out a good deal of detail; and my one regret was that I had no mask and snorkel which would have allowed me to slip overboard and watch it under water. It did not seem to have any pilot fish (which accompany most sharks) or sucker fish (*remora*) with it. Underneath it, apparently enjoying the shade provided by the vast bulk of the slow-moving monster, was another shark. It may have been 3 metres (9 feet) long, and was perhaps a grey shark, an ordinary sharp-nosed, sharp-toothed carnivore. It looked relatively tiny. The great fish swam slowly past, moving at less than an average walking pace. We again estimated that it was at least twice as long as our boat.

However, it was apparently not unaware of us, for just after it had passed it turned and swam very slowly right up to the stern. It had presumably sensed the presence of some object on the surface that should be investigated. We both stood in the stern and gazed at it; and though I felt no great apprehension I tightened my grip on the starting cord. As it approached the huge creature slowed, and very gently nosed up to us, almost, but not quite, touching the propeller. We stood there gazing down at it. The front of its head was not more than about a metre from me as I bent over to look at it better.

We could see every detail of the huge head, broader than the stern of our boat, about two metres (six feet) wide. Its enormous mouth, equipped with a myriad of small white teeth, spread in a colossal grin right round the square end of its snout. It had little eyes, situated quite high up on its head; they did not look like a normal shark's eyes, and it seemed as if it might be dim-sighted. Behind the head we could see five or six huge gill slits, each of them about one and a half metres (five feet) long, indicating the depth of the huge body. It was colossal, but apparently meant no harm, and was curious as to what this object on the surface might be.

We gazed at it, spellbound, for a long half minute. It never touched the boat at all, so that despite its huge bulk it must have been capable of delicate judgement of distance. I would guess that the tip of its nose was not more than 15 centimetres (6 inches) from the propeller; and had it given us even a gentle nudge it must have knocked us sprawling, for it undoubtedly weighed more than twenty tonnes, while our boat weighed only about a quarter of a tonne. But it did not touch us, just withdrew gently a short way, turned very slowly, and swam away, sinking almost at once into the depths. As it retreated we could see the huge grin disappearing slowly, like that of the Cheshire Cat in *Alice in Wonderland*.

We waited a few minutes before restarting the engine, hoping that it might surface again and allow us to repeat the experience. But it did not reappear. When this seemed certain we looked at each other. We were both breathless with excitement. But I and my twelve-year-old son agreed that at no time had we felt even the slightest twinge of fear, only wonder and delight that we had been privileged to see this monster at such close range, in clear, calm water. We could even appreciate that its skin must be rough, not smooth. I do not suppose that any but scuba-divers have ever seen a Whale Shark to better advantage.

No doubt, if we had rushed up to the shark in our boat, it might have been alarmed, lashed out with its tail, and flung our cockle-shell boat into the air, as harpooned basking sharks are said to do in Scottish waters. Then we might have described the encounter as a narrow shave. As it was, we looked at each other at point-blank range, and neither, I suppose, was alarmed. Certainly we were not, and felt all the time that the shark meant us no harm.

I wondered afterwards whether I had learned much about Whale Sharks from this wonderful encounter. Not much, I concluded, but something. The huge and apparently harmless beast was capable of surprisingly delicate judgement, despite the fact that it did not look as if its eyes could see the tip of its snout. Its teeth, many, very small, and pointed, evidently had some function, and perhaps it did not just swim along with its mouth wide open sieving plankton like a basking shark off the coast of Scotland. What sort of food sustained this huge bulk? I do not know, but maybe it was something that came to the surface at night in swarms, and could be gathered and minced in that great wide grin. Although it had no pilot fish (which of course may not be invariable) it acted as a shade for a more carnivorous shark, itself a big fish, but relatively appearing minute. Certainly I learned a little; but perhaps that was not important. What was important was that, unexpectedly, we had been granted, briefly, an unforgettable view of a huge creature of the sea, and that neither of us had felt any fear at all.

Further reading

THREADS OF EXPERIENCE

Darling, F. Fraser, 1939. *A Naturalist on Rona; Essays of a Biologist in Isolation*, Clarendon Press, Oxford.

1 THE ANTBEAR

Dekeyser, P. L., 1955. *Les Mammifères de l'Afrique Noire française*, 2nd edn., IFAN, Dakar.

Kingdon, J., 1971– . *East African Mammals: An Atlas of Evolution in Africa*, Academic Press, London.

2 BADGERS

Neal, E. G., 1969. *The Badger*, New Naturalist Series, Collins, London.

— 1977. *Badgers*, Blandford Mammal Series, Blandford Press, Poole.

3 BEAVERS

Rue, L. L. III, 1964. *The World of the Beaver*, Lippincott.

4 CHIMPANZEES

Goodall, J. van-Lawick, 1971. *In the Shadow of Man*, Collins, London.

5 EAGLE HILL

Brown, L. H., 1955. *Eagles*, Michael Joseph, London.

— 1976. *Eagles of the World*, David & Charles, Newton Abbot.

For scientific accounts of the Embu eagles see:

Brown, L. H., 1952–3. *Ibis*, **94**, 577–620; **95**, 74–114; **97**, 38–64 and 183–221; **108**, 531–72; **114**, 263–71.

For a classic earlier study:

Rowe, E. G., 1947. *Ibis*, **89**, 387–410 and 576–606.

6 FLAMINGOS

Allen, R. P., 1956. *The Flamingos; their Life History and Survival*, National Audubon Society, New York.

Brown, L. H., 1960. *The Mystery of the Flamingos*, Country Life, London; reproduced and enlarged, East African Publishing House, Nairobi, 1973.

For scientific accounts see:

Brown, L. H., 1958. *Ibis*, **100**, 388–420.

— and Root, A., 1971. *Ibis*, **113**, 147–72.

Brown, L. H., Powell Cotton, D., and Hopcraft, J. B. D., 1973. *Ibis*, **115**, 352–74.

For a summary of recent knowledge with full bibliography:

Kear, J. and Duplaix-Hall (eds.), 1975. *Flamingos*, T. & A. D. Poyser, Berkhampstead.

7 THE HONEY BADGER

Friedmann, H., 1955. *Honeyguides Bulletin*, **208**, Smithsonian Institution, Washington.

8 THE MOUNTAIN NYALA

Brown, L. H., 1965. *Ethiopian Episode*, Country Life, London.

Maydon, H. C., 1925. *Simen, its Heights and Abysses*, H. F. & G. Witherby, London.

For a scientific account see:

Brown, L. H., 1969. *Mammalia*, **33**, 545–97.

9 NIGHTJARS AT NIGHT

For an original account of the Cayenne Nightjar see:

Brown, L. H., 1946. *Birds and I*, Michael Joseph, London.

For scientific details:

'Nightjars' and 'Potoo', in *A New Dictionary of Birds*, ed. Sir A. Landsborough Thomson, Nelson, London, 1964.

10 OTTERS

Harris, C. J., 1968. *Otters; a Study of the Recent Lutrinae*, Weidenfeld & Nicolson, London.

Maxwell, G., 1974. *Ring of Bright Water*, Penguin Books, Harmondsworth.

11 PELICAN ISLAND

For a scientific account see:

Brown, L. H. and Urban, E. K., 1969. *Ibis*, **111**, 199–237.

For recent events at Lake Elmenteita:

Brown, L. H., Powell Cotton, D., and Hopcraft, J. B. D., 1973. *Ibis*, **115**, 352–74.

12 TIGERS

Corbett, J., 1944. *Man-Eaters of Kumaon*, Oxford University Press, Oxford.

— 1954. *More Man-Eaters of Kumaon*, Oxford University Press, Oxford.

McDougal, C., 1977. *The Face of the Tiger*, Rivington/Deutsch, London.

Perry, R., 1964. *The World of the Tiger, Cassell*, London.
Schaller, G., 1967. *The Deer and the Tiger*, Chicago University Press.

13 WHALE SHARK

Bombard, A., 1953. *The Bombard Story*, Deutsch, London.
Heyerdahl, T., 1965. *Kon-Tiki Expedition*, Allen & Unwin.

GENERAL WORKS

Dorst, J., 1970. *A Field Guide to the Larger Mammals of Africa*, Collins, London.
Encyclopaedia Britannica.
Grzimek, B. (ed.). *Animal Life Encyclopaedia*, Van Nostrand Reinhold, Wokingham.

Index

Numbers in italic refer to illustrations.

Aardvark, *see* Antbear
Abdurrahman, 121–3
Abebe, Ato Mesfin, 107–23
Ahmed, viii–xii
Akeley, Carl, 70
Alexander, Boyd, 136
Ali, Dr. Salim, 3, 168
Amadon, Dean, 19
Antbear, *xiv*, *10*, 11–17
Aonyx capensis, 141
Aplodontidae, 28
Arussi Mountains, 101, 102, 105, 120–3
Asaberu, 142–4

badgers, *xiv*, *18*, 19–25
  British, 19–25
  Honey, *xiv*, 90, 91–9
Bale Mountains, 102–5, 107–20, 123, 124–5
Bally, Peter, 75–6
Baxter, Bob, 9
beavers, *xiv*, *26*, 27–35
  American, 27
  European, 27
  Sewellel (Mountain), 27–8
Bee-eater, Carmine, 8
Berry, Hu, 168
Blue, Mohamed, ix, x
Boskovic, Z. ('Bosky'), 75–6
Boswall, Jeffery, 9, 62
Brown, Charles, 11, 185–8
Budongo Forest, 37–8
Bugoma Forest, 38–43
Burke, Victor, 152–3
Buxton, Major Ivor, 101, 102

Candy, Morag, 8
Caprimulgus ('Goatsucker'), 129
Carter, T. Donald, 101
Castor canadensis, 27
  fiber, 27
Castoridae, 28
Chimpanzee, *xiv*, *36*, 37–43
Chough, Common, 109
'Chuck-will's-widow', 130, 131; *see also* nightjars
Colorado, 32–5
Corbett, Jim, 175, 181
Corcutt, Bill, 102–5
Corophium volutator, 6–7
Crocodile, Nile, 9
Crook, John, 3

Darling, Fraser, 1, 8
Dove, White-winged, 2
Du Chaillu, 37
Dunnet, George, 7

Eagle Hill, *xiv*, 45–65
eagles, *44*, 45–65, 67
  African Hawk, 49, 51, 52, 55–6, 63
  Ayres' Hawk, *44*, 49, 51, 52, 55, 56, 57, 58, 59, 60–2, 63–4, 65
  Bateleur, 49, 51, 52, 63
  Black-breasted Snake, 49
  Brown Snake, 49, 51, 52, 53, 55, 63
  Crowned, *44*, 48, 51, 52, 55, 56, 57, 58, 59, 62, 64, 65
  Fish, 63
  Golden, 45
  Martial, 45, 48, 49–50, 55, 56, 57, 63, 64, 65
  Verreaux's, 48, 49, 51, 55, 116
  Wahlberg's, x, 49, 51, 52, 63
Eaman, Tom, 32
earth-pig, 12
Elmenteita, Lake, 81–4, 88, 166, 167–9
Embu District, 45–65
Euphorbia tree, 49, 55
Euplectes ardens, 1
Eynhus, Dan, vii–xii

Falcon, Peregrine, *44*, 60
fig-tree, 39–43
Fitzgerald, Desmond Vesey, 153
flamingos, *xiv*, *66*, 67–89
  Andean, 67
  Chilean, 81
  Greater, 76, 77, 81–4, 88, 167–9
  Lesser, 67–77, 84–8
Friedmann, Herbert, 7

Gaysay Mountain, 124
Gedligiorghis, Major Gizaw, 106–7
Gombe Stream National Park, 37
Goodall, Jane, 5, 37, 43
Guggisberg, Charles, 154

Hannington, Lake (Bogoria), 68–74, 81, 153
Hargeisa, viii–xii
Harris, C. J., 147
Hassan, Captain Mohamed, viii
Heck, François and Christine, 112–14, 124
Herodotus, 9
Höhnel, von, 8
Honey Badger, *see* badgers, Honey
honey-birds, honey-guides, 92–9
Honey guide, Black-throated or Greater, 92–9
honey-hunters, honey-men, 92–6
  Surma, 94
  Wambere, 47–8, 49, 50–1, 54–5, 56, 57, 65, 93
Hussein, 108, 118, 119
Hydrobia ulvae, 6–7

Indicator indicator, 92

Kalahari Gemsbok National Park, 96–9
Karen, 12, 56–7, 59, 177
Kicho, Njeru, 47–54, 55–6, 60–1, 77, 79, 86–7
Kruuk, Hans, 5

Lack, David, 4, 20–1
Lawick, Jane and Hugo Van, 8–9
Leakey, Richard, 37

Mackworth-Praed and Grant, 68
Magadi, Lake, 67–8, 87–8
Mary, 58
Mary II, 58
Maxwell, Gavin, 142
Maydon, H. C., 101, 102, 121
Mead, Chris, 67
Meg, 14
Mohamed, Ali Sheikh, ix, xii
Momo, 142–4
Moreau, Reg, 7
Mountain Nyala, *see* Nyala, Mountain
Mukinyu, Nyagga, 15–16, 47
Mustelidae, 91, 139

Nakuru, Lake, 85, 167, 169
Natron, Lake, 75, 76, 77–9, 85–7, 88, 154–5
Nightingale, Daphne, 169
nightjars, *xiv*, *126*, 127–37
  Abyssinian, 129
  Cayenne, *126*, 130–3
  European, 134
  Freckled, 134, 136–7
  Jungle, 129
  Long-tailed, 134, 135
  Pennant-winged, *126*, 134
  Slender-tailed, 129
  Standard-wing, *126*, 134, 135
  White-tailed, 135
North, Myles, 50, 86
Nyala, Mountain, *xiv*, *100*, 101–25
Nyctibius griseus, 130

Old Whitey, *see* William II
Onychognathus albirostris, 2
Ootacamund (Ooty), 171–2, 173
Orycteropus afer, 12
otters, *xiv*, *138*, 139–47
  Clawless, 141–4
  East Indian Clawless, 140
  Giant Brazilian, 140
  Sea, 140
owls, 120–9

Patchy, 15–16, 53, 91–2
Pelecanus conspicillatus, 150
  crispus, 150
  erythrorynchus, 150
  occidentalis, 149
  onocrotalus, 149
  philippensis, 150
  rufescens, 150
Pelican Island, 149–69
pelicans, *xiv*, *148*, 149–69
  American Brown, 149–50, 151
  American Great White, 150, 161, 169
  Australian, 150, 161
  Chilean, 149–50
  Dalmatian, 150, 160, 161
  Great White, 149, 150, 151–2, 153–69
  Pink-backed, 150, 151–3, 167
  Spotted-billed, 150, 153
Percy, William, 73
Peterson, Roger, 3
Pharaoh's Chicken, 8–9
Phoeniconaias minor, 69
Plage, Dieter, 165
Plover, Egyptian (Trochilus), 9